Excursions in geometry

C. STANLEY OGILVY

DOVER PUBLICATIONS, INC.
New York

This Dover edition, first published in 1990, is an unabridged and corrected republication of the work originally published in 1969 by Oxford University Press, New York.

Manufactured in the United States of America
Dover Publications, Inc., 31 East 2nd Street, Mineola, N.Y. 11501

Library of Congress Cataloging-in-Publication Data

Ogilvy, C. Stanley (Charles Stanley), 1913–
    Excursions in geometry / C. Stanley Ogilvy.
        p.   cm.
    Reprint. Originally published: New York : Oxford University Press, 1969.
    Includes bibliographical references and index.
    ISBN 0-486-26530-7 (pbk.)
    1. Geometry. I. Title.
QA445.O44 1990
516—dc20   90-42322
                                                                CIP

Contents

Excursions in geometry

Introduction

What is geometry? One young lady, when asked this question, answered without hesitation, "Oh, that is the subject in which we proved things." When pressed to give an example of one of the "things" proved, she was unable to do so. *Why* it was a good idea to prove things also eluded her. This girl's reactions are typical of those of a large number of people who think they have studied geometry. They forget all the subject matter, and they do not realize why the course was taught.

Forgetting the theorems is no tragedy. We forget much of the factual material we learn—or should I say encounter? —during our so-called education. Nevertheless it is a pity if a whole course is so dull that it fails to impress *any* of its content on the memories of the students. It must be admitted that traditional geometry was (and still is) guilty of this fault. Then why was it taught? Because it was supposed to present to young people a unified *logical system* on a level intelligible to them. Presumably some students got the point, but others became so involved in the details of proofs that they lost sight of the main objective.

1

The "new math" that has now been introduced in most forward-looking schools has done much toward remedying these defects. Less time is being spent on the complicated details of Euclidean (especially solid) geometry, and more on the *idea* of a geometric system. Other logical systems, much less elaborate, are being presented in order to give the student some notion of what a small forest looks like, while reducing the chance of his getting lost in the trees.

This book does not tackle these educational problems. It is not a textbook. Rather, it is intended for people who liked geometry (and perhaps even some who did not) but sensed a lack of intellectual stimulus and wondered what was missing, or felt that the play was ending just when the plot was at last beginning to become interesting.

The theorems of classical elementary geometry are nearly all too obvious to be worthy of study for their own sake. Their importance lies in the role they play in the chain of reasoning. It is regrettable that so few non-trivial theorems can be proved within the framework of the traditional geometry course when so many startlingly good ones lie just around the corner, hidden from the view of the young student. It is my purpose to present some of these to you, to recapture or reawaken your interest, with the hope that you may find that geometry is not so dull as you may have thought.

The material in this book is not all new. Most of it, though unknown to the Greeks to be sure, has been in existence for a number of years. Why, then, was it not made available to you? If there wasn't time in school, why not in college? Because you were born too late—or too soon: too late to participate in the wave of enthusiasm for geometry that swept through mathematics in the nineteenth century, when many of these things were being discovered; too soon for the "revival in geometry" now taking hold in many colleges and universities. What might be called advanced elementary geometry somehow fell from favor during the first half of the twentieth century, crowded out, probably, by the multitude of other subjects that demanded a place in the curriculum.

The question, "What is geometry?" has many answers today. There are different *kinds* of geometries: foundational, topological, non-Euclidean, *n*-dimensional, many others. We shall not attempt to investigate these. Our aim will be much more modest: to look into some of the readily accessible topics that require no formidable array of new definitions and abstractions. We shall deal mostly in the kind of geometry you already know about: the lines and points involved will be, with a few exceptions, the "ordinary" lines and points of "ordinary "geometry. We shall draw only from the kind of material that is either self-evident in the classical sense or very easy to prove. Our postulates and axioms will be those of Euclid (school geometry) unless otherwise stated, and our tools the straightedge, the compass, and a little thought.

This approach does not please the professional mathematician. He must needs start from the beginning, with a set of assumptions, and derive *everything* from these. No real mathematics can proceed along any other path. But that may have been just the trouble with your high school course: it was too formal, too cold, too bald—and hence uninspired and uninspiring. To avoid this catastrophe we will beg the forgiveness of the mathematicians, skip the formalities, and take our chances with the rest.

Of course we shall prove things. There is not much point (or fun) in being told that something is so without finding out how we know it is so. Our proofs will make liberal use of diagrams. The book is full of diagrams, and we should come to an understanding at the outset about the role that they are to play.

Have you ever seen a geometric point? Perhaps you will agree that you have not. A point has no size and "there is nothing there to see." But what about a circle? You may not be so willing to admit that you have never seen a circle, but it is very certain that you have not. A circle is defined as a set of all points in a plane equidistant from a fixed point. Already we have guaranteed its invisibility: if a point has no thickness, neither has a "row" of them, and there is still nothing there to see. What you *do* see when you draw a circle with a compass is only an attempt to picture a circle, and a poor attempt at that. It is not a circle because: (1) It is not made up of points strung out along a line: the alleged "line" has width. (2) Even if the line (or rather its picture) were to be made microscopically thin, accurate measurement would detect unequal distances from the center—assuming that the "center" could be located in any meaningful way, which it could not. (3) The alleged circle does not lie in a plane; a piece of paper is very far from a true geometric plane. (4) Even if the paper *were* a plane, the ink has thickness building up away from the paper. And so on.

Have you ever seen *any* geometric figures? Certainly not. They are defined in such a way that they can never have physical or tangible existence. It is a tribute to our powers of mental abstraction that even though none of us has ever seen points or lines, we can talk about them with confidence. When I say "straight line," you have no trouble visualizing exactly what I mean. It is in this inner realm of mental visions that all geometry really takes place—*not* on the paper. One must guard against thinking, "The diagram proves it." Appearances are often misleading; diagrams are useful only as an aid for picturing things that can (at least theoretically) be stated *and proved* without them. Yet they are so useful in clarifying our thinking that only the most abstract purists attempt to dispense with them entirely.

The abstract nature of geometry was at least partially understood and appreciated by the Greeks. That is why the "permissible tools" of classical geometry are straightedge and compass only. Consider the problem of trisecting an angle, which has no solution with these tools alone. Then why not use a protractor? Just measure the angle, divide the number of degrees by three, and there you are. But *where* are you? This superficial solution to the problem disturbs our feeling of what is acceptable and proper in geometric society, so to speak. It is the very *purity* of the ruler and compass that suits them so well to the purity (abstractness) of the subject. If your sensibilities are outraged by the idea of measuring the number of degrees in an angle, you have already taken a long stride into geometry.

It would be well to mention that the Notes in the back of the book contain not only source references but also much other material. You should consider the Notes as a running commentary accompanying the text; they may occasionally help you over the rough spots. Check them from time to time to see whether there is anything you have missed.

Are you ready for a mathematical ramble? Then let's go.

I

A bit of background

A PRACTICAL PROBLEM

The owner of a drive-in theater has been professionally informed on the optimum angle θ (theta) that the screen AB should present to the viewer (Fig. 1). But only one

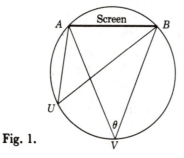

Fig. 1.

customer can sit in the preferred spot V, directly in front of the screen. The owner is interested in locating other points, U, from which the screen subtends the same angle θ.

The answer is the circle passing through the three points A, B, and V and the reason is:

Theorem 1. An angle inscribed in a circle is measured by half the intercepted arc.

Inasmuch as $\angle AUB$ is measured by the same arc as $\angle AVB$, the angle at U is the same as the angle at V. A proof of the theorem is given in the Notes in case you have forgotten it.

A typical example of a calculus problem is this: Of all triangles on the same base and with the same vertex angle, which has the greatest area? We are able to steal the show briefly from the calculus and solve this problem almost at a glance, with the aid of Theorem 1. If AB is the base and the given vertex angle is θ, then all the triangles have their vertices lying on the circle of Fig. 1 (turn it upside down if you prefer the base to be at the bottom). But

$$\text{area} = \tfrac{1}{2}\ \text{base} \times \text{altitude}.$$

One-half the base is a constant; and the altitude (and hence the area) is greatest for AVB, the isosceles triangle.

Even to set the problem up for the calculus is awkward, and several lines of careful calculations are required to solve it.

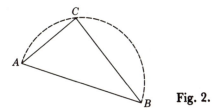

Fig. 2.

From Theorem 1 we also have the useful corollary that any angle inscribed in a semicircle is a right angle (measured by half the 180° arc of the semicircle). If by any chance we can show that some angle ACB (Fig. 2) is a right angle, then we know that a semicircle on AB as a diameter must pass through C. This is the converse of the corollary.

A BASIC THEOREM

If two chords of a circle intersect anywhere, at any angle, what can be said about the segments of each cut off by the other? The data seem to be too few for any conclusion; yet an important and far-reaching theorem can be formulated from only this meager amount of "given":

Theorem 2. If two chords intersect, the product of the segments of the one equals the product of the segments of the other.

This theorem says that in Fig. 3A, $PA \cdot PB = PC \cdot PD$. (The dot is the symbol for algebraic multiplication.) What do we mean when we talk about "the product of two line segments"? (There is a way to "multiply two lines" with ruler and compass, but we won't have to do it.) What the theorem means is that the products of the respective *lengths* of the segments are equal. Whenever we put a line segment like PA into an equation, we shall mean the *length* of PA. (Although we shall soon have to come to grips with the idea of a "negative length" for the present we make no distinction between the length of PA and the length of AP: they are the same positive number.)

The Greek geometers took great pains to enumerate the different "cases" of each theorem. Today we prefer, when possible, to treat all variants together in one compact theorem. In Fig. 3A the chords intersect inside the circle, in 3B outside the circle, and in 3C one of the chords has become a tangent. Theorem 2 holds in all three cases, and the proofs are so much alike that one proof virtually goes for all.

You may object, and say that in Fig. 3B the chords don't intersect. But they do when extended, and in this book we

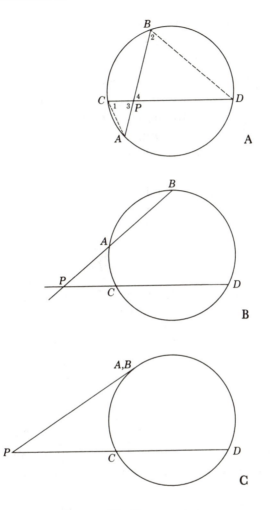

Fig. 3.

shall say that one line intersects a second when it in fact
only intersects the *extension* of the second line. This is just
part of a terminology, in quite general use today, that may
be slightly more free-wheeling than what you have been

accustomed to. In the same vein, P divides AB *internally* in Fig. 3A, whereas in Fig. 3B P is said to divide the chord AB *externally,* and we still talk about the two *segments* PA and PB.

To prove Theorem 2 we need two construction lines, indicated in Fig. 3A. Perhaps you should draw them also in Figs. 3B and 3C, and, following the proof letter for letter, see for yourself how few changes are required to complete the proofs for those diagrams. In Fig. 3A, $\angle 1 = \angle 2$ because they are inscribed in the same circular arc (Theorem 1), and $\angle 3 = \angle 4$ (why?). Therefore triangles PCA and PBD are similar, and hence have proportional sides:

$$\frac{PA}{PD} = \frac{PC}{PB}$$

or $$PA \cdot PB = PC \cdot PD.$$

MEANS

The average of two lengths (or numbers) is called their *arithmetic mean*: $\frac{1}{2}(a+b)$ is the arithmetic mean of a and b. The *geometric mean* is the square root of their product: $\sqrt{ab}$. The arithmetic mean of 8 and 2 is 5; their geometric mean is 4.

It is easy to find the geometric mean of two positive numbers algebraically by solving for x the equation $a/x = x/b$; for this says $x^2 = ab$. For this reason the geometric mean is also called the *mean proportional* between a and b.

Is there any way to find these quantities with ruler and compass? The arithmetic mean is easy enough: simply lay off the two lengths end to end on the same straight line and bisect the total segment. A method for the geometric mean is suggested by the equation

$$\frac{AB}{x} = \frac{x}{BC}.$$

Using $AB+BC$ as diameter, draw a circle and the chord perpendicular to that diameter at B (Fig. 4). This chord is bisected by the diameter, and we recognize a special case of Theorem 2, namely, $x \cdot x = AB \cdot BC$, which says that x is the mean proportional between AB and BC.

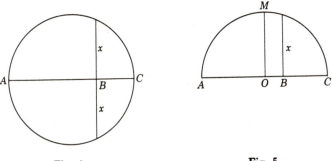

Fig. 4.   Fig. 5.

In practice one uses only half the figure (Fig. 5).

Inasmuch as MO, the perpendicular from O, the midpoint of AC, is the *longest* perpendicular to the semicircle from its diameter and is also a radius $= \frac{1}{2}AC = \frac{1}{2}(AB+BC)$, we have:

Theorem 3. The geometric mean of two unequal positive numbers is always less than their arithmetic mean.

Why did we need the word "unequal"?

We now restate the part of Theorem 2 that applies to Fig. 3C in a form that may look familiar:

Theorem 4. If, from an external point, a tangent and a secant are drawn, the tangent is the mean proportional between the whole secant and its external segment.

Of course the two segments *PA* and *PB* become the same length when the chord is moved into the limiting position of tangency, bringing *A* and *B* together. If you are bothered by this, the proof (see Notes) is quite independent of any limiting process.

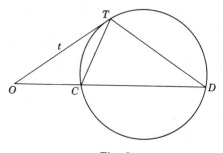

Fig. 6.

If we relabel the diagram as in Fig. 6 and then call *OT* by the single letter *t*, we have

$$t^2 = OC \cdot OD$$

if and only if *t* is a tangent. These labels are chosen to fit the lettering of the next chapter.

I trust you will agree that we have done nothing difficult so far. Yet you may soon be surprised at the structure we are about to build on this small but sturdy foundation. We are now ready to look at some of the geometry they didn't teach you.

2

Harmonic division and Apollonian circles

HARMONIC CONJUGATES

Is it possible to divide a line segment internally and externally in the same ratio? In Fig. 7, do points C and D exist such that

$$\frac{AC}{CB} = \frac{AD}{BD}?$$

The answer is yes; C and D are then said to divide AB *harmonically*.

Fig. 7.

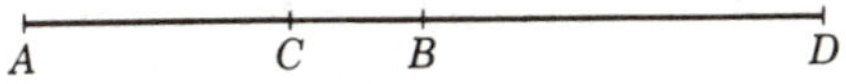

In fact there are infinitely many solutions to the problem. We shall find that for any point C between A and B there is a specific point D that satisfies the requirement. This is another way of saying that one can find two points dividing the segment harmonically in any required ratio.

If two points C and D divide AB harmonically, then, from the last equation, $AC \cdot BD = CB \cdot AD$. Dividing both sides by BD and AD, we get

$$\frac{CB}{BD} = \frac{AC}{AD}.$$

We state this as:

Theorem 5. If AB is divided harmonically by C and D, then CD is divided harmonically by A and B.

The points C and D are called *harmonic conjugates* of each other with respect to A and B.

THE CIRCLE OF APOLLONIUS

Suppose that we want to find points C and D that divide AB harmonically in some given ratio, k.

The Greek mathematician Apollonius discovered that a circle may be defined in a way quite different from the usual "all points equidistant from a fixed point." Apollonius' definition states that if a point moves in such a way that its distance from one fixed point is always a constant multiple of its distance from another fixed point, then its path is a circle. The proof goes as follows:

Suppose A and B are two fixed points. The moving point, P, can be anywhere, provided that $AP = k \cdot BP$, where the constant k is any positive number except, for the moment, 1. That is, k may be less than or greater than 1. Given only the points A and B and a number k, you could always find a suitable point P, although you might make one or two false starts. With any radius, draw an arc with center at B; with radius k times as long, draw an arc with center at A. If these arcs intersect, their intersection is a suitable P; if they do not intersect, re-adjust the first radius and try again. Because of this trial-and-error procedure, it is not the best way to divide the segment harmonically. We shall do better shortly.

We assume now that we have found, or have been given, a point P such that

$$\frac{AP}{BP} = k.$$

In Fig. 8, draw PC, the bisector of the interior angle APB of

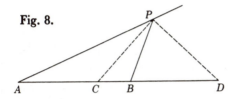

Fig. 8.

triangle APB, and PD, the bisector of the exterior angle at P. From plane geometry we have:

Theorem 6. The bisector of any angle of a triangle divides the opposite side into parts proportional to the adjacent sides.

$$\frac{AC}{CB} = \frac{AP}{BP} = k. \tag{1}$$

Inasmuch as this is equally true of exterior angle bisectors, we also have

$$\frac{AD}{BD} = \frac{AP}{BP} = k. \tag{2}$$

Combining equations (1) and (2) yields

$$\frac{AC}{CB} = \frac{AD}{BD} = k, \tag{3}$$

and our first problem is solved: we have divided the segment AB harmonically in the ratio k.

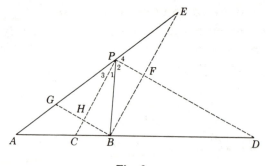

Fig. 9.

Because we pulled Theorem 6 (especially equation 2) rather abruptly out of the hat, we might well slow down and prove it. In Fig. 9, $\angle 1 = \angle 3$ and $\angle 2 = \angle 4$ (given). Adding equals to equals, $\angle 1 + \angle 2 = \angle 3 + \angle 4$. But the whole totals 180°, so $\angle 1 + \angle 2 = 90°$, a right angle. Draw BE parallel to CP (and hence perpendicular to PD). The two right triangles PFE and PFB are congruent (why?), making $PE = PB$. Furthermore, the two parallel lines cut off the following proportional segments:

$$\frac{AC}{CB} = \frac{AP}{PE} = \frac{AP}{BP}. \tag{1}$$

The proof goes the same way for

$$\frac{AD}{BD} = \frac{AP}{GP} = \frac{AP}{BP}. \tag{2}$$

Returning now to equation (3), we note that it is *independent of the letter P*. This means that we could do the same proof over again with a new point P, and so long as we used the same k we would arrive at the same points C and D. For *every* such point P, $\angle CPD$ is a right angle, and we observed in Chapter 1 that when this happens P lies on a semicircle whose diameter is *CD*. If *P* is below the line *AB* we get the rest of the Apollonian circle of Fig. 10. Note that *B* is not the center.

Suppose a ship leaves point *B* and steams in a fixed direction at a constant speed. A second ship, leaving *A* at the same time, can go *k* times as fast as the first ship. Assuming a plane ocean, what course should the fast ship steer in order to intercept the slow ship as quickly as possible? We have just solved the problem. If the navigator plots the Apollonian circle of the two points *A* and *B* for the constant *k*, he need only extend the slow ship's course until it cuts the circle, say at *Q* (Fig. 10), and then head for that point. $AQ = k \cdot BQ$, and they must arrive simultaneously at *Q*.

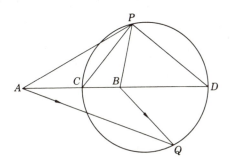

Fig. 10.

COAXIAL FAMILIES

Let us see what happens when *k* is changed in the harmonic division problem. For each value of *k* we get a new pair of points *C* and *D* and hence a new Apollonian circle for the points *A* and *B*. As *k* approaches 1, the circles become larger. If *k* = 1, the "circle" is a straight line, the perpendicular bisector of segment *AB*. You may consider it a circle of infinitely long radius if you like. Our diagrams so far have all indicated that k is greater than 1, written $k > 1$. For $k < 1$ (k less than 1), the circle appears on the other side of the perpendicular bisector, and the point D of the harmonic division appears to the left of A. Clearly $k = 1$ is a special case. Where does D go when C is midway between A and B?

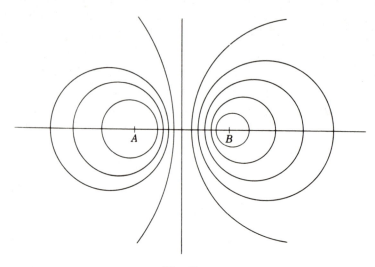

Fig. 11.

Figure 11 pictures the Apollonian circles corresponding to various values of k. It is called a non-intersecting coaxial family.

Two circles (any two curves, for that matter) are said to intersect *orthogonally* if their respective tangents at the point of intersection form a right angle there (Fig. 12). We hope that the following facts are obvious:

Theorem 7. (1) If one intersection of two circles is orthogonal so is the other; (2) two circles are orthogonal if, and only if, the radius of either to the point of intersection is

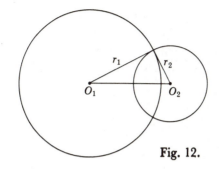

Fig. 12.

tangent to the other (because a tangent is always perpendicular to the radius at the point of contact); (3) two circles are orthogonal if, and only if, $r_1{}^2 + r_2{}^2 = \overline{O_1 O_2}^2$ (by the Pythagorean theorem).

Given a harmonic division of AB by C and D; if O is the midpoint of AB, let $OA = OB$ be called r for reasons that will immediately become apparent. Then (see Fig. 13),

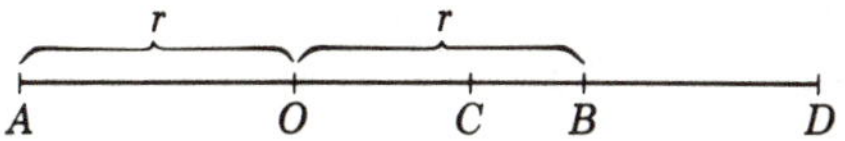

Fig. 13.

$$\frac{AC}{CB} = \frac{AD}{BD}$$

can be rewritten

$$\frac{r + OC}{r - OC} = \frac{OD + r}{OD - r}.$$

If you multiply all this out and collect the pieces, you should end up with

$$r^2 = OC \cdot OD.$$

This last equation, then, is *equivalent* to the statement that the division is harmonic.

Now draw circle α (alpha) so that *AB* is its diameter, and let β (beta) be *any* circle passing through *C* and *D*. These two circles must intersect (twice). Let *T* be a point of intersection, and draw *OT* (Fig. 14). We have just decided that

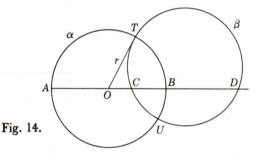

Fig. 14.

$r^2 = OC \cdot OD$. But this can happen if, and only if, *r* is tangent to β (by the last part of Theorem 4). Therefore, by Theorem 7 (2), the two circles are orthogonal. We restate this as:

Theorem 8. If a diameter of one circle is divided harmonically by another circle, the two circles are orthogonal.

The converse is also true, as is easily proved by taking the steps in the reverse order.

From Theorem 8 we derive a number of remarkable consequences. Circle α is just one of a coaxial family, all members of which have diameters divided harmonically by the *same* points *C* and *D*. Hence circle β cuts all the members of the α family orthogonally. Not only that: there is nothing special about β, except that it must pass through *C* and *D*; but infinitely many circles do that, and they all behave like β insofar as they all are orthogonal to every member of the α family. The centers of the β family lie on the perpendicular bisector of *CD* (why?). The β's form an intersecting coaxial family. Fig. 15 shows some members of both families. Every intersection of an α by a β is orthogonal.

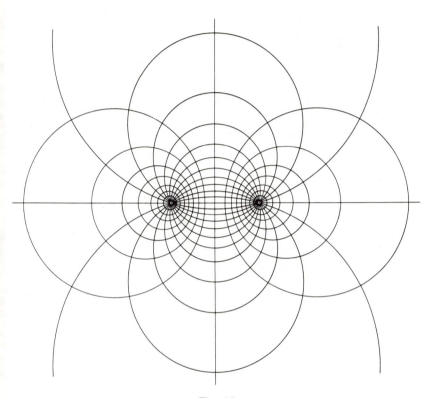

Fig. 15.

Consider any one β circle and two given α circles. The center of the β circle is a point from which the pairs of tangents to the two α circles are equal, because all four tangent lines are radii of the β circle. From the center of another β circle the same thing is true. Thus the line of centers of the β family is the locus of points from which equal tangents can be drawn to the two given α circles. It is called the *radical axis* of those two circles. And now if we substitute any other member of the α family for one of the two given α circles, leaving the other one as before, we see that all members of the α family share the same radical axis: the line of centers of the β family.

If two given circles intersect, what is their radical axis? Simply interchange the letters α and β everywhere in the last paragraph and note that all statements and conclusions still hold. The radical axis of two intersecting circles is therefore their common chord.

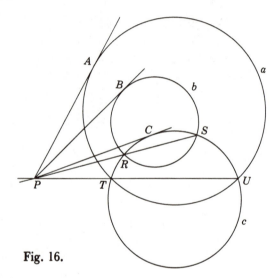

Fig. 16.

If two given circles do not intersect, we know all about their radical axis in theory; but how would you go about actually finding it with ruler and compass? Draw any third circle c that intersects the given circles a and b, at whatever angle (Fig. 16). Then P, the intersection of the common chords, lies on the radical axis. Another random circle (not shown) intersecting a and b similarly produces a second point, say Q; and PQ is the radical axis.

Do you see why this is so? From P draw tangent lines to A, B, and C. Because TU is a *common* chord, we can apply Theorem 4 first to circle a and then to circle c to obtain

$$PA^2 = PT \cdot PU = PC^2.$$

Likewise, since RS is a chord common to b and c,

$$PB^2 = PR \cdot PS = PC^2.$$

But these equalities show that $PA^2 = PB^2$, or $PA = PB$, which makes P a point on the radical axis. One other such point, Q, is all we need to determine the straight line.

The proof is considerably abbreviated in the following more high-powered presentation: TU is the radical axis of a and c, and RS is the radical axis of b and c. Hence P, their intersection, is a point from which the tangents to *all three*— a, b, and c—are equal, by two applications of the definition of the radical axis for two intersecting circles.

3
Inversive geometry

Perhaps you did calculations "by logarithms" somewhere in your high school career. Why? Logarithms are exponents, to some base (in school the base was 10). In the process of *multiplying* numbers with like bases, exponents are *added*. It is easier to add than to multiply. This is only one kind of calculation that might be easier "by logs," but it will do for an illustration. The two numbers to be multiplied could be of four or five digits each. It is a fairly simple matter to look up each number in the log table, write down its logarithm, add them up, and then find the "antilog" of the result.

What is actually going on in such a process depends on the fact that there is a *one-to-one correspondence* (no overlapping in either direction) between the positive real numbers and their logarithms. The log table actually *lists* this correspondence. You should think of it as a transformation:

$$y = \log x.$$

Every x has just one logarithm, that is, just one y, and vice

24

versa. By looking up the logs, we *transform* the problem from the positive number level to the logarithm level, where it is easier to solve. When it has been solved (for instance, after the addition is done), then we transform the answer back to the original level through the same one-to-one correspondence furnished by the table.

To reiterate: an operation, or problem, was difficult at the given level; we transformed the whole thing to another, easier level, solved it there, and transformed the solution back to the first level.

The point is that although the transformation was one-to-one, the operation performed on the second level was different from the one that would have had to be done on the original level. This is the only advantage of the method. Our logarithm example is not very spectacular, but it is relatively familiar. The solution of differential equations by the Laplace transform is another, well known to engineers and physicists. We are about to investigate a third instance of the method.

INVERSION

Can you turn a circle inside out? If it *could* be done, all the points that were previously on the inside would now be on the outside, and vice versa. There are many different ways in which such an *inversion* can be accomplished geometrically. We shall select one particular way because it is tidy and produces good results.

Given a circle of radius r, let each point C inside the circle be transformed to a point D outside the circle in such a way that the distances comply with

$$OC \cdot OD = r^2 \tag{1}$$

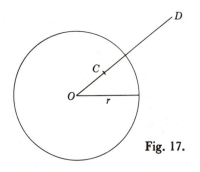

Fig. 17.

and the line OC passes through D (Fig. 17). C and D are called inverse points with respect to the circle of inversion. We can delete the "outside" and "inside" phrases from the definition: equation (1) automatically takes care of that. Also note that all points previously outside end up inside: the inverse of D is C. The points C and D simply swap places.

This then is a one-to-one transformation of the plane onto itself. Each point (except which one?) has a definite place to go under the transformation, and there is no confusion of two points trying to occupy the same place. Such a transformation is often called a *mapping*.

What good is it? We shall soon see that it enables us to solve some hard geometric problems easily. In the meantime, it has other incidental attractions.

Where have you seen equation (1) before? It is exactly the condition (page 19) guaranteeing that C and D divide harmonically the diameter on which they lie. So an equivalent definition of inversion could be stated in terms of harmonic conjugates. (State it.)

Since two inverse points are nothing but harmonic conjugates with respect to the ends of the diameter, we already know how to find the inverse of a given point. But our method of Chapter 2 was clumsy and we promised then to smooth it out.

Method I: Given a point P and a circle, draw the diameter AB through P and connect A and B to any other point Q on the circle (Fig. 18). Now construct $\angle 2$ such that $\angle 2 = \angle 1$. Then the dotted side of $\angle 2$ cuts AB in P', the inverse of

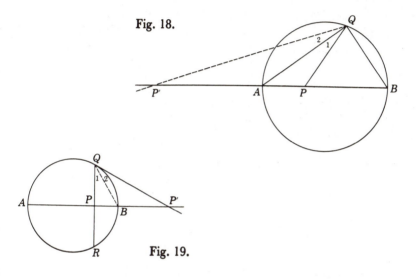

Fig. 18.

Fig. 19.

P. Why? Because AQB is a right triangle (inscribed in a semicircle), and we have exactly the situation of Figs. 8 and 9. Hence we know that A and B divide PP' harmonically; therefore, by Theorem 5, we know that P, P' divide AB harmonically. This construction works equally well if we start with P' and have to find P.

We note in passing that for a circle with $P'P$ as diameter, A and B are inverse points.

Method II: If P is a point inside the circle, draw PQ perpendicular to AB (Fig. 19). From Q draw a tangent; P' is the point where this intersects AB.

Proof IIa: If this claim is correct, then it must be true that
$\angle 2 = \angle 1$ (by the previous construction, Method I). But
angles 1 and 2 subtend equal arcs QB and BR. (*Q.E.D.*).
Proof IIb: Draw a circle on QP' as a diameter (Fig. 20).
It must pass through P, because QPP' is a right angle by
construction. But the tangent to one circle is the radius of
the other, making the circles orthogonal (Theorem 7 (2));
and hence, by the converse of Theorem 8, AB is divided
harmonically by P, P'. To a mathematician, IIb is perhaps
more "elegant" than IIa. It is surely more sophisticated.

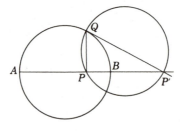

Fig. 20.

So far, Method II works only if P is inside the circle. Or
does it? If P is outside, the problem reduces to finding the
tangent to a circle from an external point. There are many
ways to do this, but pivoting the ruler through P until it just
touches the circle is *not* one of them. This might be a slap-
dash practical method, but it is not an accepted Euclidean
construction. We can use our ruler only to draw a line
determined by two points. How, then to find Q? Possibly the
simplest way is to draw a semicircle on diameter OP (the
dashed semicircle of Fig. 21). Because it is inscribed in a
semicircle, $\angle OQP$ is a right angle. This makes Q the required

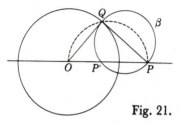

Fig. 21.

point; for PQ, being orthogonal to a radius at the point of contact, must therefore be the tangent from P. We need only to draw a circle β on PQ as diameter, and β cuts OP at P', the inverse of P.

Method III: If P is outside the circle, draw an arc with P as center and PO as radius, cutting the circle at S (Fig. 22).

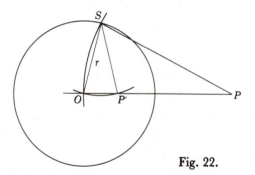

Fig. 22.

An arc of radius r with S as center will now cut OP at P'. Proof: Isosceles triangles SOP' and POS share a base angle, and hence are similar. Therefore,

$$\frac{OP'}{r} = \frac{r}{OP}, \qquad \text{Q.E.D.}$$

If P is inside the circle, start with center at P and radius r

to find S. Then the perpendicular bisector of OS passes through P'. The details are left to you.

There are still other methods for finding inverse points.

We hope you have noticed that there seems to be an exceptional point in any inversion. What happens to the center, O? Points very near O go very far away, and, if the situation is to be continuous, O itself must in some sense "go to infinity." Therefore in order to keep our correspondence one-to-one, without any exceptional points, we must postulate just *one* point at infinity. Is that too strange? If you like to visualize models, think of the diagram being drawn on the surface of a very large sphere (the earth, for instance). Then the role of the point at infinity could be played by the point diametrically opposite to O on the surface of the sphere, as if the whole plane were wrapped around the sphere and "buttoned up" at the opposite side. But that is only a device, and if you don't like it, forget it. The inversive plane is not actually meant to be wrapped around anything. Nevertheless you can see that it does differ somehow from the (ordinary) plane of elementary plane geometry.

You may object, "But it doesn't make sense to say that there is only one point at infinity. I know very well that if I start in one direction and you start in another, we will arrive at infinity at two different points."

I'm sorry, but you know nothing of the kind. You actually have no idea what will "happen" at infinity; furthermore, you can never "arrive" there to find out. These phrases are meaningless. The more we think about it, the more we realize that we know nothing intuitively about infinity. We must simply put aside all talk about what "must be true" there. *Nothing* is "true" there in any sense that can be proved from scratch. Note that the famous parallel postulate of Euclid is a *postulate*, not a theorem that can be derived from the other postulates. We can postulate anything we please, anywhere in mathematics, provided we then faithfully and obediently take the consequences. It turns out that in some geometries it is convenient to postulate a whole line of points at infinity, but in inversive geometry we choose to postulate exactly one point at infinity, namely, the inverse of O. This happens to be entirely consistent (it introduces no self-contradictions), which is always a test of whether a postulate is a useful one.

INVARIANTS

Suppose we have some geometric figures, like triangles or circles, drawn on a plane; and suppose we invert that plane with respect to some fixed circle. It is natural to ask, what happens to the geometric figures? Are they only made larger or smaller, or do they also change shape? Some distortion must take place; how much and what kind of distortion? What properties of the original figures remain unchanged? We say that the properties that remain unchanged are "preserved under inversion," and call them *invariants* of the transformation.

Certainly size is not an invariant. Inasmuch as the inside of the circle has to be spread over the whole of the plane outside the circle, things must "get bigger" (or smaller if they were previously outside and are brought inside). Distance is not preserved.

With the possible exception of the center of the circle of inversion, points *near* each other, in some definable sense, before inversion, turn up near each other after inversion. "Neighborhoods" are preserved. They may become smaller or larger, but they are preserved. The transformation does not break up a small connected area into several disconnected areas. It is possible to formalize this situation mathematically (we shall not do so), and the theory is abbreviated into the statement that the transformation is *continuous*.

The demand that a transformation be continuous is by no means a minor requirement. Its strength can be glimpsed through the following elementary problem.

Suppose a piece of string is just long enough to be stretched straight across a room, from wall *A* to the opposite wall *B*. It is then used, without being cut, to tie up a parcel of complicated shape, so that the string becomes wound into many loops and knots around the parcel. The tied package is left in the middle of the room. Is there now any part (point) of the string that is located at exactly the same distance from wall *A* that it was when it was stretched across the room? The answer is "Yes, there must be"; and the proof requires the application of an apparently trivial theorem about continuity. Even with these hints, you may move to the head of the class if you can prove it before looking into the Notes.

The continuity of the inversion transformation might lead us to think that perhaps *similarity* is preserved. Is a triangle expanded (or contracted) into a similar triangle by the transformation? The answer is no. Then what does happen to a triangle? We start by answering the simpler question, what happens to a straight line?

As a consequence of the original definition of inversion, a straight line through the center remains a straight line. But it is worth noting that, although a straight line through the center is invariant, the individual points on the line are not: they all swap places with other points, except which two? The ones *on* the circle of inversion stay put. Thus the circle of inversion itself is invariant both ways—as a circle, and also pointwise.

What about a straight line not through the center? To

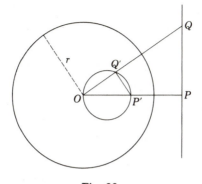

Fig. 23.

begin with we can take one that does not cut the circle, like the line in Fig. 23 on which P and Q are located. Draw OP, the perpendicular from O to the line. Draw OQ, where Q is any other point on the line. Then if P' and Q' are the inverses of P and Q, we have

$$OQ \cdot OQ' = r^2 \quad \text{and} \quad OP \cdot OP' = r^2.$$

Together these two equations say

$$OP \cdot OP' = OQ \cdot OQ' \quad \text{or} \quad \frac{OQ'}{OP'} = \frac{OP}{OQ}.$$

But if the two triangles OPQ and $OQ'P'$ (note the order of the vertices) have a common angle and the adjacent sides proportional, they are similar. Therefore $\angle OQ'P'$ is a right angle. But Q is *any* point on the straight line; so, for various positions of Q, Q' will trace out a locus such that $\angle OQ'P'$ is always a right angle. This locus is a circle, on OP' as diameter, by the corollary to Theorem 1. We have discovered something quite unexpected: A straight line *not* through the center of inversion inverts to a circle *through* the center of inversion.

A convenient name for the configuration into which a figure is cast by the inversion transformation is its *image*. The image of the straight line is the circle. The same transformation interchanges the image points with the original points, so that the converse is automatically true: the image of a circle through the center is a straight line not through the center. The size of the circle is connected with the distance from the center of inversion to the straight line by the equation $OP \cdot OP' = r^2$.

What if the given straight line cuts the circle of inversion in two points? Then those two points are invariant, but the rest of the proof holds exactly as before. The result is a diagram like Fig. 24. The size of the circle is still governed by $OP \cdot OP' = r^2$; but it is much easier to locate the circle by simply passing it through the points A, B, and O of Fig. 24.

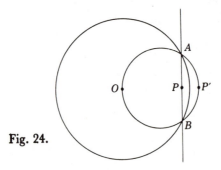

Fig. 24.

We wrap this up as a single theorem:

Theorem 9. The image of any straight line not through the center of inversion is a circle through the center of inversion and, conversely, the image of a circle through the center is a straight line not through the center.

The next question is, what is the image of a circle that does not pass through the center of inversion? The answer is given by

Theorem 10. The image of a circle not passing through the center of inversion is another circle not passing through the center of inversion.

The proof is like the previous one, only slightly more elaborate. Starting with a circle outside the circle of inversion, draw OQP so that PQ is a diameter of the given circle. (Fig. 25). Let P', Q' be the inverses of P, Q, and let R' be the in-

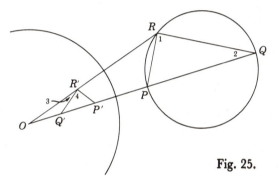

Fig. 25.

verse of R, any other point on the given circle. Note that this time we cannot immediately conclude that $\angle 4$ is a right angle, although in fact it is, for we do not have triangles PQR and $P'R'Q'$ similar. What we do have is

$$OP \cdot OP' = OR \cdot OR'$$

(because both of these are equal to r^2).

Hence
$$\frac{OP}{OR} = \frac{OR'}{OP'}$$

Since they have the angle at O in common, this makes triangles OPR and $OR'P'$ similar. In exactly the same way triangles OQR and $OR'Q'$ are similar.

From the first two,

$$\angle OPR = \angle OR'P'.$$

But $\angle OPR$, an exterior angle of triangle PRQ, is equal to the sum of the two opposite interior angles. Therefore we can substitute this sum for $\angle OPR$:

$$\angle 1 + \angle 2 = \angle OR'P'.$$

From the second pair of similar triangles:

$$\angle 2 = \angle 3.$$

Substituting, $\angle 1 + \angle 3 = \angle OR'P'.$

But $\angle 4 + \angle 3 = \angle OR'P'.$

Hence $\angle 4 = \angle 1.$

But $\angle 1$ is a right angle, being inscribed in a semicircle. Therefore $\angle 4$ is likewise a right angle, and thus the locus of R' is another semicircle, on $Q'P'$ as diameter, for any pair of positions (R, R'). This is what we set out to prove.

You can satisfy yourself, by trying it, that the proof goes the same way if the given circle is inside of or cuts across the circle of inversion. It is indeed surprising that, except for those covered by Theorem 9, circles invert into circles. We can call these circles invariants if we mean only that their "circularity" and not their size is preserved.

Are the respective *centers* preserved? That is, does the center of the circle invert to the center of the image circle? I put it to you to try to show that it does not.

Our new knowledge sheds further light on Fig. 14, where the diameter AB is divided harmonically by C and D. In the language of the present chapter, C and D are inverse points with respect to circle α. Now what is the image of the whole circle β? It must be another circle β', which must pass through the invariant point T, and through the two points C and D because they are mutual inverses. But it is possible to draw only *one* circle through three given points. This means that β' is the same circle as β; its points are not invariant (they swap

positions), but it is invariant as a configuration. This is a
further confirmation of the converse of Theorem 8: *any
diameter of α is cut by β in inverse points.*

Observe that it is only orthogonal circles that behave this
way: if β cut α at any other angle, points T and U would be
preserved under inversion but C and D would not exchange
places, and β′ would be a circle different from β.

We are closing in on the answer to our original question,
"What is the image of a triangle under inversion?" If no side
of the triangle lies along a radial line (through O), then
each side, being a segment of a straight line, must invert to
an arc of a circle; and we have for the image some sort of
curvilinear triangle. Can we say anything about the angles of
this curvilinear figure? Yes: they are the same as the angles
of the original triangle. *Angle is invariant* under inversion.

What do we mean by the angle between two curves? It is
defined as the angle between the tangents to the curves at the

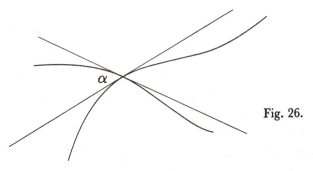

Fig. 26.

point of intersection (Fig. 26). There are of course two possible angles (actually four), but we can fix attention on the
acute angle α if there is one (they could all be right angles).
The two curves (of any kind, not necessarily circles) invert to
two new curves, and these image curves, though quite different from their parent curves, will cross at the same angle α.

Any transformation that accomplishes this is called *conformal*. We shall see that inversion is in fact *anti-conformal*: the size of α is preserved, but its sense (direction) is reversed.

Theorem 11. Inversion is anti-conformal.

We shall prove first that the angle between a curve *s* and a radial line (through *O*) is preserved. We have (Fig. 27) a

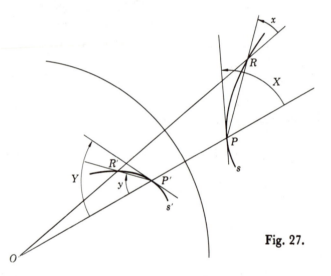

Fig. 27.

radial line cutting the curve *s* at *P*, and cutting the image curve *s'* at *P'*. We know already (see the proof of Theorem 10) that from the similarity of their respective triangles, $\angle ORP = \angle O'P'R'$. Therefore $\angle y = \angle x$. But now let *R* approach *P* along the curve *s*. Then *R'* approaches *P'* along *s'*. As the line *PR* approaches tangency to *s* at *P*, the line *P'R'* approaches tangency to *s'* at *P'*. Thus as $\angle x$ approaches $\angle X$, $\angle y$ approaches $\angle Y$; and inasmuch as $x = y$ all along the way, their limits are equal: $\angle X = \angle Y$.

If two curves intersect, say at *P*, we need only draw the radial line *OP* and apply the proof to the second curve. It is then a matter of equals added to (or subtracted from) equals, for the total angle between the two curves.

We have now completed the answer to the original question. A triangle inverts to a curvilinear triangle (about whose area we have said nothing); its sides become arcs of circles; and its angles are preserved in size but reversed in sense.

CROSS-RATIO

Turn back now to Fig. 7 (p. 13), which depicts the harmonic division of a line segment by two points. If direction is taken into account, we can indicate positive lengths as being measured from left to right and negative lengths from right to left. Thus the negative of CB is BC, and we have

$$\frac{AC}{BC} = -\frac{AD}{BD}.$$

One of the four lengths has been reversed, the other three remaining the same, so that there is one change in sign. The above equation says that

$$\frac{AC}{BC} \Big/ \frac{AD}{BD} = -1$$

The left-hand side is a "ratio of ratios," that is, the ratio of the two pieces that divide the line internally divided by the ratio of the two pieces that divide the line externally. Because these two ratios are equal numerically, we have the quotient 1, or, with due regard to sign, -1.

What if the quotient is *not* -1? That is, given C, suppose D is the wrong point for a harmonic division. The quantity

$$\frac{AC}{BC} \Big/ \frac{AD}{BD}$$

still has a name, the *cross-ratio*, even though its value is not -1. When the cross-ratio is -1, the division is harmonic; otherwise it is not.

We re-emphasize that *any* four points A, C, B, D (note the order) on a line determine a cross-ratio as defined above. It pervades several branches of geometry. One cannot "see" the cross-ratio at a glance, nor predict its behavior, but it is a remarkably persistent quantity under transformation. Among other things, it survives inversion: cross-ratio is an inversive invariant.

If four points on a line undergo an inversion, they usually end up not on a straight line but on a circle. Can we define cross-ratio on a circle? Yes, but not yet. So we prove only:

Theorem 12. The cross-ratio of four points on a radial line is preserved under inversion.

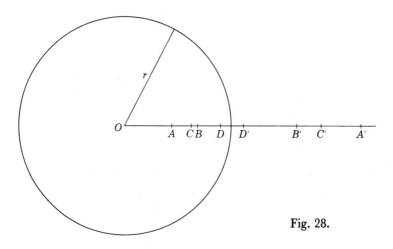

Fig. 28.

In Fig. 28, let OA be abbreviated as a, $OC = c$, and so on. Then

$$a' = \frac{r^2}{a}, \qquad b' = \frac{r^2}{b}, \qquad \text{etc.}$$

$$AC = OC - OA = c - a \qquad \text{in this notation.}$$

CB would be $OB - OC = b - c$. But we need that segment in the negative direction, $c-b$. The whole cross-ratio is then

$$\frac{AC}{BC} \div \frac{AD}{BD} = \frac{c-a}{c-b} \div \frac{d-a}{d-b}.$$

What happens under inversion? All four image points appear in the opposite order, so all pairs are reversed and their cross-ratio is

$$\frac{a'-c'}{b'-c'} \div \frac{a'-d'}{b'-d'} = \frac{\dfrac{r^2}{a}-\dfrac{r^2}{c}}{\dfrac{r^2}{b}-\dfrac{r^2}{c}} \div \frac{\dfrac{r^2}{a}-\dfrac{r^2}{d}}{\dfrac{r^2}{b}-\dfrac{r^2}{d}}$$

The r's cancel, and collecting the fractions gives

$$\frac{\dfrac{c-a}{ac}}{\dfrac{c-b}{bc}} \div \frac{\dfrac{d-a}{ad}}{\dfrac{d-b}{db}} = \frac{c-a}{c-b}\cdot\frac{b}{a} \div \frac{d-a}{d-b}\cdot\frac{b}{a} = \frac{c-a}{c-b} \div \frac{d-a}{d-b}.$$

We summarize the things that are invariant under inversion. This list is not complete, but it contains all those that we have dealt with.

1. The circle of inversion is pointwise invariant: *each point* on it remains fixed. All the rest are invariant, but not pointwise.

2. Radial lines.

3. Circles orthogonal to the circle of inversion. Each of these is a fixed circle, inverting into itself.

4. Circles not through the origin. (These are invariant only in the sense that they go into other circles not through the origin; only their circularity is preserved.)

5. Angle (but reversed in sense).

6. Cross-ratio of points on a radial line.

4
Applications of inversive geometry

Consider the family of circles tangent to the Y axis (and hence to each other) at the origin (Fig. 29). Let circle C, not tangent, pass through the origin also. How does C intersect all the circles of the family?

One way to answer the question is to perform an inversion with respect to a circle—any circle—centered at the origin. Because all the circles of the problem pass through the origin, *all* of them will become straight lines (Theorem 9). Furthermore, since no member of the family intersects another member, their image curves are *parallel* straight lines. The image of C is a straight line, which of course intersects every one of the parallels at the same angle. But angle is preserved (Theorem 11). Therefore the circle C, before inversion, intersected all the circles of the family at the same angle.

This solution is typical of the method. A suitable inversion transformed the problem into a much easier problem—so easy, in fact, that it was not even necessary to draw the

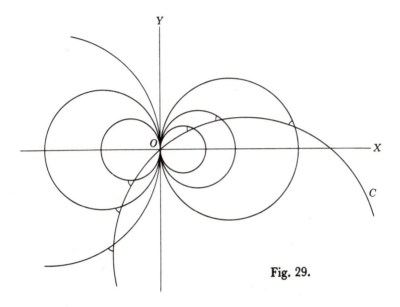

Fig. 29.

transformed diagram. It was only necessary to know that it consisted of a bunch of parallels cut by a transversal. From the transformed diagram we could read off the solution to the original problem, making use of one of the invariant properties.

Perhaps you have already noticed that this problem can easily be solved without any inversion. Fix attention on *one* circle of the family and circle C (Fig. 30). When any two circles intersect, their angles of intersection are equal at both points. One can prove this by reference to the symmetry across the line of centers. $\angle\alpha = \angle\beta$, and hence $\angle\alpha = \angle\alpha'$. But the same argument holds for *any* member of the family, each of which shares $\angle\alpha$ with C. Hence all angles of intersection equaling α, equal each other. I hope that you share with me the feeling that this is not as "snappy" as the inversion proof.

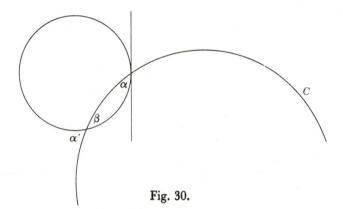

Fig. 30.

The problems that follow are not so readily solved. Some of them are quite difficult without the aid of inversive geometry. But it is perfectly true that they all can be done without inversion. Inversive geometry is not a magic wand that does *unsolvable* geometric problems; no such wand exists. What it does do is in line with what all mathematics does: when judiciously applied, it makes hard things easy, and brings to light previously hidden properties.

Let three circles have a point in common. Suppose that the common chord of two of them is a diameter of the third. If this is true for *two* of the three common chords, then we can prove that it must be true for *all three* of them.

We select this as a second illustration because it is just a little harder to do without inversion than the last example was. You might try it: but be on your guard against assuming too much from the visual evidence of the diagram. We shall now prove it by an inversion.

In Fig. 31 we are given that AO and BO are diameters, and we must prove that CO is also. We invert the diagram with respect to O. (Henceforth we shall use this language as an abbreviation for "with respect to a circle whose center

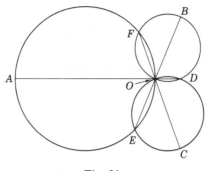

Fig. 31.

is at O," whenever we are not concerned about the size of the circle of inversion; that is, any such circle will do.) The straight lines through O are invariant; and the circles become straight lines not through O. Intersections are preserved,

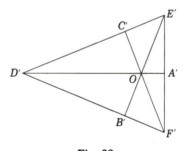

Fig. 32.

so that our inverted diagram must look something like Fig. 32, where each primed letter labels the point that is the image of that point designated in Fig. 31 by the same letter unprimed: A goes to A', etc.

Now the angles at A' and B' are right angles, because the diameters intersected the circles at right angles at A and B, and angle is preserved. Hence $A'D'$ and $B'E'$ are *altitudes* of the triangle; and because the three altitudes of any triangle are always concurrent, the line $F'OC'$ has to be also an altitude, and therefore orthogonal to $D'E'$. But this proves the theorem, because before inversion FC must therefore have been orthogonal to the third circle, and hence a diameter of that circle.

PEAUCELLIER'S LINKAGE

During the nineteenth century considerable interest centered in the problem of translating rotary motion into linear motion by means of a mechanical linkage. For a time it was thought that it was not possible to do this without cams or pistons or like devices requiring indirect thrusts and frictional losses; but eventually (in 1864) the problem was solved by Peaucellier, whose linkage makes use of the fact that the image of a circle through the center of inversion is a straight line.

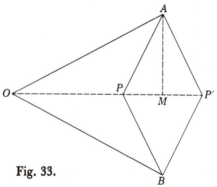

Fig. 33.

The linkage is diagrammed in Fig. 33. The four bars forming the diamond are of equal length. The two longer bars are equal to each other, $OA = OB$. The bars are pivoted at O, A, B, P, and P'. At O there is a fixed pivot (attached, say, to the drawing board); otherwise the apparatus is free to move in the plane. But regardless of its position,

$$OP \cdot OP' = (OM - PM)\,(OM + PM)$$
$$= OM^2 - PM^2$$
$$= (OA^2 - AM^2) - (AP^2 - AM^2)$$
$$= OA^2 - AP^2,$$

a *constant* inasmuch as OA and AP are bars of the linkage. But

$$OP \cdot OP' = \text{constant}$$

is all that is necessary for P' to be the image of P with respect to some circle (we don't care what circle), whose center is at O. If, then, we could somehow make P move along another circle that passed through O, P' would move along a straight line, the image of that circle.

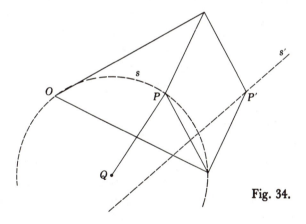

Fig. 34.

But this is readily accomplished. It is only necessary to add a seventh bar PQ of length equal to OQ for any fixed point Q (Fig. 34). Anchor the end of this bar at Q, so that when the bar turns it carries P along circle s that passes through O. Then P' must move along the image of s, that is, the straight line s'.

You can easily construct this linkage with soda straws or cardboard strips and pins. You will find, however, that it is not possible to carry the point P all the way around the circle s: the linkage will jam at a certain point. It is said that some reciprocating ventilating machinery did actually operate for many years in the British Houses of Parliament with a Peaucellier linkage as an integral part of the mechanism. The problem lost its practical interest with the introduction of superior lubricants and precision bearings that greatly increased the efficiency of the ordinary cam and piston.

APOLLONIUS' PROBLEM

A problem that has come down to us from antiquity is Apollonius' problem of the three circles: to construct, with ruler and compass, a circle tangent to three given circles. The given circles are of different sizes, they may or may not intersect one another, etc. Of course there are impossible situations. If one circle is wholly inside another no solution is possible; if the three have collinear centers and the same radii, the solution "circles" would be a pair of parallel tangents; I leave it to you to envisage other exceptional cases. In general, however, there are eight solutions: the circle we seek may be externally tangent to all three given circles; it may be internally tangent to all three (surround them all); it may surround two and not the third (in three ways); or it may surround one and not the other two (in three ways).

This is one of the problems whose solution is a heavy task without the aid of inversion. We shall solve the first case; the others are managed in similar fashion.

Given the three circles of Fig. 35, we first resort to a

simplifying device that would eventually occur to anyone attacking this problem by whatever method. We increase the radii of all three circles by the same amount, δ, just sufficient to make the two nearest ones tangent to each other. It is evident that if we can solve the Apollonius problem for the three new circles, the *center* of the solution circle will be the

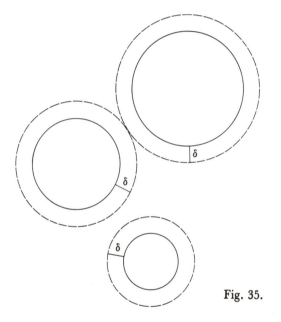

Fig. 35.

same for the original problem, and the *radius* of the solution circle will differ from the required radius by δ. So we turn our attention to the simplified version, redrawn in Fig. 36.

Let us now perform an inversion with respect to the dashed circle centered at O. Any circle would do; but since we are free to choose, we use for the circle of inversion the one that leaves the third circle C invariant. You know how to do this: the circle of inversion must meet C orthogonally, and we have described the method of drawing it (see Fig. 21, page 29).

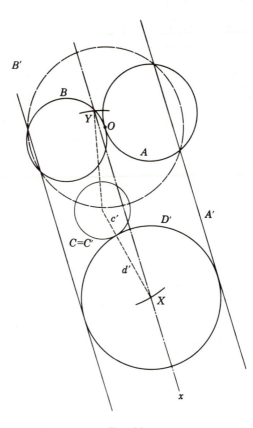

Fig. 36.

The resulting inverted diagram is so simple that it can be drawn on the same figure, labeled with the primed letters.

The next step is to find a circle D' tangent to A', B', and C'. But that is easy because A' and B' are parallel (why?); hence its radius d' is half the distance between the parallel lines, and its center lies on the midline x. If c' is the radius of C', then an arc with radius $(c'+d')$ must cut x at the center of D'. There are *two* such points, X and Y; what is the significance of the second one?

It remains only to find the circle D, the pre-image of D', and the problem is solved because of the preservation of tangency by inversion. It is an easy matter, following the methods of Chapter 3, to find this image. It will be the circle required for the simplified Apollonius diagram, and the ultimate solution is found by adding δ to its radius.

STEINER CHAINS

What happens if we invert the whole of Fig. 15 (page 21) with respect to point C? All the β circles pass through C and hence become straight lines not through C; but they must all intersect at another point, the image of D. The α circles, on the other hand, must become new circles orthogonal to these straight lines. Thus we have Fig. 37, in which the concentric circles are the images of the α circles, the

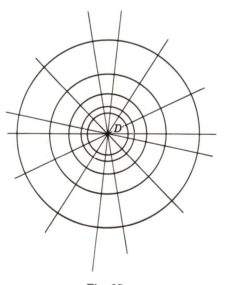

Fig. 37.

radial lines are the images of the β circles, and D' is the image of D. (The image of C is of course the point at infinity.)

Theorem 13. Given two non-intersecting circles, it is always posssible to invert them into a pair of concentric circles.

Proof: Draw the radical axis of the two circles (see page 22), and call them two α circles. Using any two points on this line as centers, draw two circles orthogonal to the α circles (by the procedure of Fig. 21). These are then β circles and so must intersect at C. Now we showed in the previous paragraph that inverting the figure with respect to C sends the two α circles into two concentric circles.

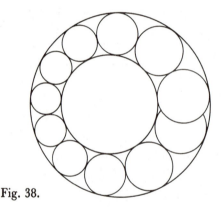

Fig. 38.

We are now in a position to investigate a configuration called Steiner's chain of circles, after the accomplished nineteenth-century geometer. Suppose we are given two circles, one inside the other. We can insert a chain of circles tangent to both these circles and each tangent to its two neighbors. In general, the inserted chain will not "come out even," but will look something like Fig. 38. But if various conditions of size and spacing happen to be just right, the chain may close, as in Fig. 39.

Suppose we are lucky enough to find a pair of "base" circles with the closed chain of Fig. 39. You have probably guessed that shifting one member of the chain, however slightly, would upset this delicate arrangement; in other words, if we started a new chain with a circle different from any of the eight shown, the new chain would not close.

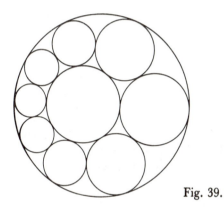

Fig. 39.

Steiner showed that this guess is wrong: *it makes no difference where we start the chain.* The remarkable fact is that, for any two given circles, if one closed chain is possible, then a chain started from *any* position will close. This property is immediately revealed if we invert Fig. 39 so that the two given circles become concentric circles, by Theorem 13. Then the Steiner chain becomes a chain of *equal* circles contained between them (Fig. 40). These equal circles can be moved around (visualize ball bearings), and from an infinite number of different positions they can be returned by inversion to an infinite number of different closed Steiner chains in Fig. 39.

There is a Steiner chain even if more than one round is required to complete it. If there were exactly "ten and one

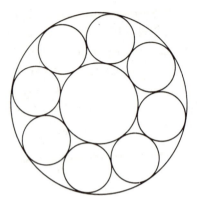

Fig. 40.

half" circles (see Notes) in Fig. 38, they would come out even the second time around for a 21-circle chain of two windings. If, however, the number of circles in one circuit is irrational, then the chain never closes.

THE ARBELOS

Fig. 41.

The configuration of Fig. 41, consisting of three semicircles tangent as shown and with diameters along a common line, is called the arbelos, or the shoemaker's knife.

If we inscribe a chain of circles as in Fig. 42, we find that they have the following unexpected property: the vertical

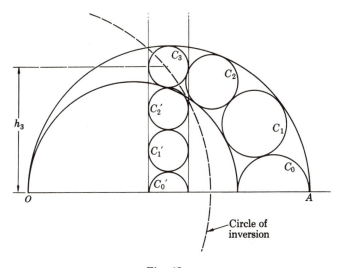

Fig. 42.

distance h_n from the line OA to the center c_n of the nth circle C_n is exactly nd_n, where d_n is the diameter of C_n. This is proved by inverting in a circle centered at O and orthogonal to C_n. The diagram shows the inversion done for $n = 3$. The two large semicircles of the arbelos invert to parallel lines tangent to C_3 as shown. C_3 inverts into itself, and C_2, C_1, and the semicircle C_0 must invert to C_2', C_1', and C_0' with tangency duly preserved throughout. It follows that $h_3 = 3d_3$.

5
The hexlet

An *ellipse* can be defined as the path of a point that moves (in a plane) so that the sum of its distances from two fixed points remains constant. Put two pins in a sheet of paper on a drawing board. Tie a piece of string, one end to each pin, so that the string is not stretched tight but is longer than the distance between the pins. Let your pencil take up the slack and, restraining it by the string, draw a curve. The curve is an ellipse (**Fig. 43**), because $d_1 + d_2 =$ constant, the fixed length of the string. The points F_1 and F_2, where the pins are located, are called its foci.

A *hyperbola* is similarly defined, except that the word *sum* is replaced by *difference*, so that $d_1 - d_2 =$ constant. (Query: in Fig. 43, how is the other branch obtained?)

A *parabola* is the path of a point that moves so that its distance from a fixed point is equal to its distance from a fixed line.

These three curves (and some special cases of them) are called the *conic sections*. We shall see later how they can be

56

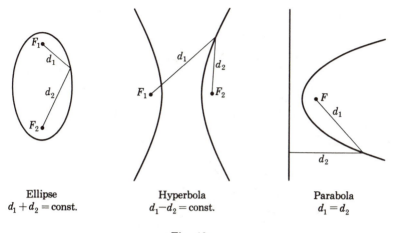

Ellipse	Hyperbola	Parabola
$d_1 + d_2 = \text{const.}$	$d_1 - d_2 = \text{const.}$	$d_1 = d_2$

Fig. 43.

obtained as the intersection of a plane and a cone. Their algebraic equations are all of the second degree, making it easy to study them analytically. At the moment we need only a theorem that follows from the definitions.

A PROPERTY OF CHAINS

Theorem 14. The centers of the circles of a Steiner chain lie on an ellipse.

Proof: Given the fixed circles of Fig. 44, with centers O_1, O_2, and radii r_1, r_2, let r be the radius of a *variable* circle whose center is P, tangent to the other two as shown. External tangency at T_1 means that the line of centers PO_1 consists of two radii; internal tangency at T_2 means that P lies on a radius r_2. That is,

$$PO_1 + PO_2 = (r_1 + r) + (r_2 - r) = r_1 + r_2 = \text{constant}$$

for all positions of P. This is exactly the definition of an ellipse with foci at O_1 and O_2.

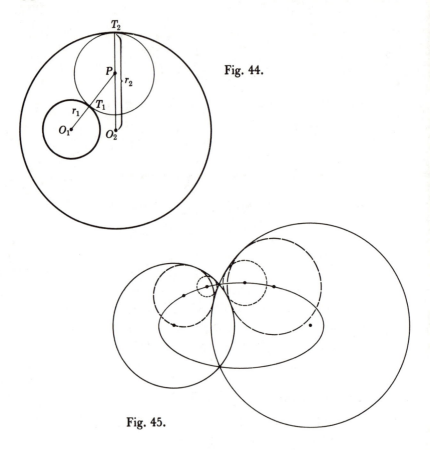

Fig. 44.

Fig. 45.

Theorem 14 applies also if the two base circles are such that the chain does not close (Fig. 38) or if the two base circles are internally tangent (Fig. 42), because the proof uses only the tangency at T_1 and T_2. In fact, the two base circles can intersect each other in two real points (Fig. 45); but now the roles of r_1 and r_2 are interchanged as P moves through the intersection.

What happens if the two base circles are entirely outside each other? There are still infinitely many circles tangent externally to one base circle and internally to the other (Fig. 46). What is the locus of their centers?

$$PO_2 - PO_1 = (r+r_2) - (r-r_1) = r_2 + r_1 = \text{constant.}$$

This time it was necessary to take the *difference* of two distances to obtain a constant—the definition of a hyperbola with foci at O_1, O_2.

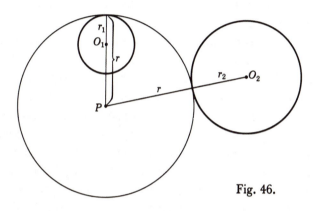

Fig. 46.

If the variable circle is externally tangent to both fixed circles (I leave it to you to draw the picture), the equation is

$$PO_2 - PO_1 = (r+r_2) - (r+r_1) = r_2 - r_1,$$

a different constant; so we are now on a different hyperbola. For any one conic it is necessary to settle temporarily on *one* of the four possible kinds of tangency: external to both O_1 and O_2, internal to both, external to O_1 but internal to O_2, external to O_2 but internal to O_1. With this understanding we can collect all the cases in one basket and call it

Theorem 14 Generalized. The centers of all circles tangent to two fixed circles lie on a conic.

We have produced ellipses and hyperbolas. Question: Is there a parabolic case?

SODDY'S HEXLET

What is the three-dimensional analog of Apollonius' problem of the three circles? Shifting from circles to spheres, we notice that the extra dimension provides an infinity of spheres, each tangent to three given spheres. Is it in general possible to find a sphere tangent to each of *four* given spheres A, B, C, and D? We use the alternative attack mentioned in the Notes: shrink the radii of all four spheres by an amount equal to the radius of the smallest sphere, say D, thus reducing D to a point, and then perform a 3-space inversion with respect to that point. We state without proof the obvious analogies: A sphere of inversion is used instead of a circle of inversion; spheres become spheres, except for those spheres through the center of inversion, which become planes not through the center of inversion. The inversion sends the three spheres A, B, and C into three new spheres A', B', and C'. Consider a plane tangent to A', B', and C' (there are eight such planes possible.) Generally this plane will not pass through D, so that if we return the spheres to their original positions by another inversion in D, the plane becomes a fifth sphere, E, passing through D and tangent to A, B, and C. If we now restore the radii to their original sizes and reduce the radius of E accordingly, the problem is solved.

The analog of Steiner's chain is called Soddy's Hexlet, a necklace of spheres, all tangent to three given spheres A, B, and C and each tangent to two neighbors in the chain. We start by selecting *any* sphere S_1 that is tangent to A, B, and C. The last paragraph tells us that we can then find a second sphere S_2 tangent to A, B, C, and S_1, a third tangent to A, B, C, and S_2, and so on. But will S_n, the "last" sphere, be tangent to S_1 to complete the necklace? You will, of course, think that in general this does not happen. But unlike the Steiner circles, and quite astonishingly, not only does the chain always close perfectly, but it always contains exactly six spheres! Hence the name *hexlet*.

The simplest hexlet is best visualized by taking A and B much larger than C. A hole remains between these three tangent spheres. The hexlet girdles C and threads the hole.

The proof of the hexlet property is remarkably easy. If we invert the figure consisting of the four tangent spheres A, B, C, and S_1, with respect to the point of tangency of A and B, then A and B become parallel planes (cf. Apollonius' problem), and the image of C is a sphere C' contained between these planes and tangent to both. S_1 was tangent to all three, so it becomes a sphere the same size as C', lying between the planes and also tangent to C'. Note that all this happens regardless of the original size of S_1. Now exactly five more spheres equal in size can be added in a closed chain around C', their centers lying on the vertices of a regular hexagon. Try it by placing six golf balls around a seventh on the table (or even six pennies around a seventh). When the figure is inverted back to the original orientation, these six become the hexlet.

Professor Frederick Soddy of Oxford discovered this configuration in 1936 the hard way, without the aid of inversive geometry, and published much information about the relative sizes of those particular hexlets whose spheres all have radii that are *reciprocals of integers*. Fig. 47 attempts to depict such a hexlet. A and B are taken to be spheres of diameter 1, radius $\frac{1}{2}$, each tangent to the plane of the paper at point O, one sphere located above the page, the other beneath the page. The dotted circle shows their outline, looking down along their line of centers, which is perpendicular to the page. Circle C has radius $\frac{1}{11}$. The other six circles, with radii as indicated, represent spheres of the hexlet, and all of them, together with C, have their centers in the plane of the paper. Soddy has shown, among many other things, that these dimensions make the six spheres tangent to A and B, as well as to C.

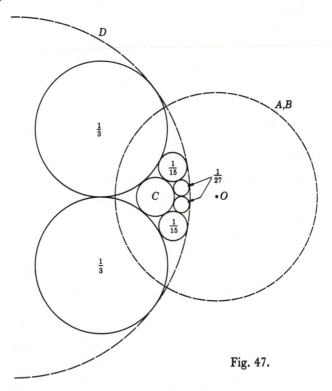

Fig. 47.

From this point on, we shall consider the hexlet to be six spheres of a "fluid" chain, without fixed radii. Thus Fig. 47 depicts just one of an infinite number of positions of six (necessarily varying) spheres as they roll around C, in the same way that any Steiner chain has an infinite number of positions. Each sphere of the hexlet is always tangent to two neighboring ones, and all six are always tangent to A, B, and C, which we shall call the *fixed spheres*. The symmetrical model obtained by inversion with respect to O, the point of tangency of A and B, will be referred to as the inverted hexlet or the inverted model.

The points of mutual tangency of the six equal spheres of the inverted hexlet lie on a circle orthogonal to all six spheres at these points of contact. This circle inverts to a circle, and orthogonality and tangency are preserved. It follows that, in the original hexlet, (1) the six points of tangency also lie on a circle, and (2) the *centers* of the six hexlet spheres lie in a plane. Since nothing has been said in this paragraph about the relative sizes of A, B, and C, it is true of *any* hexlet that its centers lie in a plane.

In the inverted model there is another sphere, D', concentric with C' but surrounding the hexlet and tangent to all six spheres. In the particular case of Fig. 47, where the inversion preserves the central plane (the plane of the paper), D' is the image of D, a sphere again surrounding the hexlet, partial cross-section of which is shown dashed in the figure. Thus Theorem 14 tells us that the centers of this special hexlet lie on an ellipse with foci at the centers of C and D. We shall soon find that even if A and B are not the same size, the centers of a Soddy hexlet always lie on an ellipse.

Figure 47 indicates that, although we can start with an S_1 anywhere in the flowing chain, the size of S_1 would seem to be limited. Certainly S_1 must not be so small that it could fall freely through the hole between A, B, and C without touching them. This puts a lower limit on the diameter of S_1. It is evident that, for the particular trio A, B, and C of Fig. 47, there is also an upper limit. Sphere C is nested between A and B in such a way that too large an S_1 could not touch C, being blocked by A and B. These limits are the smallest and largest diameters of the circles of the Steiner chain on their trip around the two fixed circles C and D.

SOME NEW HEXLETS

Consider now a new trio of fixed spheres with a larger C. Suppose, in fact, that C is big enough to project *beyond* the outline of A and B: "No hexlet," said Professor Soddy. He envisioned this difficulty, and referred to the limiting case approached as C approached the size at which a plane would

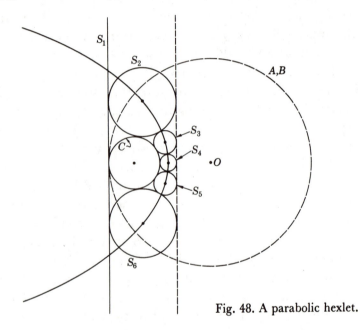

Fig. 48. A parabolic hexlet.

be tangent to A, B, and C (Fig. 48). As far as he was concerned, you could have a hexlet only if the three fixed spheres were suitably chosen to begin with. But how can this be? Our original inversion proof does not depend on the sizes of A, B, and C: in the inverted model there is *always* a hexlet.

Let us try to make a hexlet with C "too big," as in Fig. 49. For simplicity we continue to take A and B to be the same size and tangent at O as in Fig. 47, although the discussion is essentially unchanged when A, B, C are all of different radii. The mutual points of contact of the six spheres of Fig. 49 still lie obediently on a circle. We'll come to the locus of

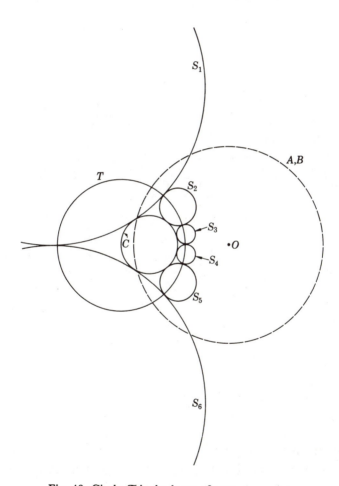

Fig. 49. Circle T is the locus of tangency points.

their centers presently. If we try to make them flow around
C, say counterclockwise, S_1 cannot go very far before becom-
ing a *plane*, and then *reversing its concavity* (Fig. 50). Later it

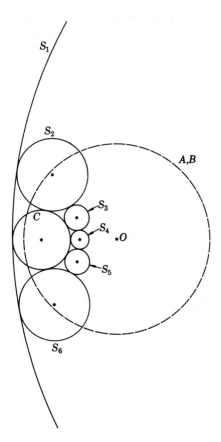

Fig. 50.

passes through the plane stage again and flips back to the
original concavity. Each sphere has to be turned inside out,
so to speak, twice on every round trip.

It is clear why Soddy did not consider this case. He was a physical scientist, and he thought (and wrote) of his spheres as "beads of a necklace." No necklace ever behaved so wildly. But it turns out that these overlooked hexlets are just the ones that possess certain intriguing geometric properties.

How can all this happen, when it does not happen in the inverted model where they flow smoothly around like balls in a ball-race? The answer is that a sphere of the hexlet becomes a plane, and then reverses concavity, whenever its image *passes through the center of inversion.* In a hexlet of Soddy's type, C is of such a size that the whole inverted model never touches O. We are free to choose for our sphere of inversion any sphere centered at O; let us choose the one orthogonal to C, so that C inverts into itself. Then the critical radius for C is seen to be one-quarter of the radius of A (or B); for then each sphere of the inverted hexlet will be just big enough to touch O as it passes. This is Soddy's limiting case; the inverse of a sphere as it brushes past the center of inversion is momentarily the (left-hand) vertical plane of Fig. 48, after which it returns to a sphere of the same concavity. If, on the other hand, C is even larger, then each sphere of the inverted model has to surround O for a time, during which period its original is concave the other way.

What is the locus of centers of this new type of hexlet? If A and B are not the same size we cannot fall back on the earlier method. All we know is that the centers lie in a plane. Consider just two fixed spheres of different sizes, tangent externally, together with a variable sphere tangent to them both but with its center always in one plane. We know from the generalized version of Theorem 14 that the locus of the center of the variable sphere is a hyperbola with foci at the centers of the two fixed spheres. It follows that if we now remove the restriction that the center of the variable sphere must lie in a plane, its locus becomes a *hyperboloid of revolution* with the same foci as before. Thus, regardless of the third fixed sphere, all spheres of the hexlet must have their centers on this hyperboloid. In addition, they lie in a plane. But the plane section of a hyperboloid of revolution is a conic.

In Soddy's hexlet this conic is an ellipse, in the limiting case it is a parabola, and in the curvature reversing case it is a hyperbola. It is convenient to refer to a hexlet as elliptic, parabolic, or hyperbolic, accordingly.

It is instructive to sketch these conics. In Fig. 48 the centers have to recede in only one direction indefinitely far, and then return from the same direction. This is characteristic of the way in which the parabola tends to infinity. But in the hyperbolic case a center recedes to infinity along a branch of the hyperbola in the direction of an asymptote, it is "at infinity" when the sphere in question becomes a plane, and then it comes back on the *other branch* of the hyperbola, staying on that branch as long as the sphere is concave in the opposite sense. Thus the center of any one sphere completes a trip around the entire conic as its image sphere makes one revolution around C'.

An elliptic hexlet has the center of inversion O always completely exterior to every sphere of the inverted model. Hence there exist two spheres, each tangent to all six spheres of the model and passing through O. The pre-images of these spheres are planes not through O, but tangent to the original hexlet. Thus a hexlet of elliptic type fits into the dihedral "V" between two tangent planes. If we increase the size of C so that the elliptic hexlet approaches a parabolic one, these two planes approach coincidence; and in the parabolic case they do coincide, in the position of the dashed vertical line of Fig. 48. In the image model of this situation, the two surrounding tangent spheres have merged into one through O. There can be no such sphere or spheres in the hyperbolic case.

Every inverted model hexlet has an infinite number of spheres surrounding it and tangent to it. Their centers are collinear, lying on that perpendicular to the central plane which passes through the center of C'. All those not passing through O invert back to spheres tangent to the original hexlet, and in the elliptic case one of them plays the role of the outer fixed Steiner circle.

Now let us consider hexlets around the same A and B spheres but with increasingly large C spheres. A, B, and C are of course fixed for any one hexlet; so when we speak of C as a variable sphere we are asking, What is the effect on the resulting hexlet when the size of C is changed? As the radius of C tends toward infinity, so that C tends toward a plane, then the image C' moves toward O, passing through O when C actually is a plane. As C' continues to move, now with O inside it, C comes back as a sphere with the opposite concavity, *surrounding* the six spheres of the hexlet and A and B. C is now the analog of the outer Steiner circle. It is interesting to note that this was Soddy's starting point. He called it "the simplest case of the hexlet." In Fig. 51, C has radius $= 1$, A and B, tangent to the page above and below at O and shown dotted, each has radius $= \frac{1}{2}$, and the six equal spheres of the hexlet have radius $= \frac{1}{3}$.

We remarked earlier that the six points of mutual tangency of any hexlet lie on a circle. There is one case where that statement fails unless we allow a straight line as a special case of a circle: when the circle on which the tangency points lie in the inverted model *passes through* O. Then the preimage of that circle is a straight line, on which lie the points of mutual tangency and the six centers of the hexlet spheres.

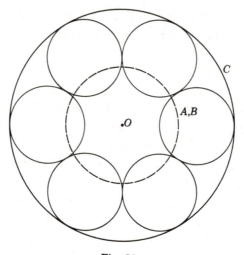

Fig. 51.

This occurs if and only if the three fixed spheres A, B, and C are all the same size. Let us follow the behavior of a sphere S_1 as it oscillates back and forth to complete a "circuit" of this special straight hexlet.

S_1 is small as it passes through the hole between the three equal spheres, and increases to infinite radius as it becomes a plane tangent to them. When it is a plane its model image is passing through O. Moreover, O is now exactly at a point of tangency with the neighboring model sphere, say S_6'. Therefore, at this instant S_6 is also a plane tangent to A, B, and C, as indicated in Fig. 52A. As S_1 proceeds on its way, it reverses its concavity and now surrounds all the other eight spheres, tangent internally to A, B, C, S_2, and S_6 (Fig. 52B). As S_2 continues to push upward and grow, S_6 shrinks, and S_1 approaches and attains the tangent plane position held by S_6 in Fig. 52A. It then reverses concavity again and shrinks, to pass through the hole and begin another round trip. If you trace it carefully you will find

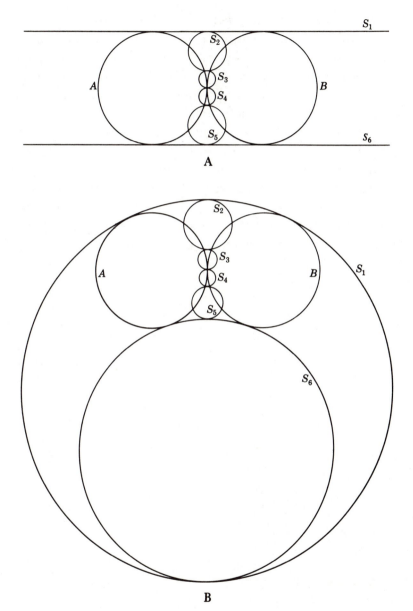

A

B

Fig. 52.

that the center has traversed the whole infinite straight line
from one "end" to the other twice in one complete circuit.
This sounds much like the action in the hyperbolic case;
and indeed the line of centers is called a degenerate hyperbola, the limiting case where the hyperbola has collapsed
into two coincident straight lines.

There is surely more to be said about the hexlet, and further
phenomena remain to be discovered. I hope you are tempted
to do some investigating on your own. Here is an almost
unexplored corner of geometry where you can find your way
with very little mathematical preparation and come upon
some gems well worth the search—gems not large, perhaps,
but of excellent quality and a pleasing luster.

6

The conic sections

We shall now investigate some further properties of the three curves defined in the opening section of Chapter 5.

THE REFLECTION PROPERTY

Theorem 15(A). The locus of the center of a variable circle, tangent to a fixed circle and passing through a fixed point inside that circle, is an ellipse.

This is a corollary of Theorem 14 if r_1 is allowed to shrink to zero. Or we can easily prove it independently: In Fig. 53, V is the variable circle, F is the fixed circle and O_1 is the fixed point. Then

$$PO_1 + PO_2 = r + PO_2 = TO_2 = \text{constant}$$

for all positions of P.

From a piece of paper cut a circular disk (five or six inches in diameter will do), and mark a point O_1 anywhere on the disk except at its center. Fold the paper so as to bring a point on its circumference into coincidence with O_1.

Unfold, and repeat the process several times using different circumference points. You will find that the creases envelop the ellipse of Theorem 15(A).

To see why this occurs, fix attention on some one circumference point, say T of Fig. 53. O_1 is a second point on V; and the only way to bring two circumference points of the same circle into coincidence is to fold along the circle's diameter. Hence the crease goes through P. Furthermore, it is tangent to the ellipse at P; for if it were not, it would cut the ellipse again at some other point P', which would have to be the center of *another* circle tangent at T and passing through O_1, an impossibility.

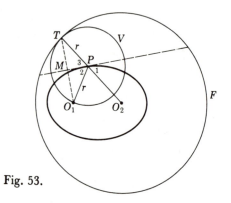

Fig. 53.

Triangles O_1PM and TPM have three sides respectively equal, and hence they are congruent. Therefore $\angle 2 = \angle 3$. But of course $\angle 1 = \angle 3$, being vertical angles; so $\angle 1 = \angle 2$, and we have discovered a property of the ellipse: the focal radii to a point make equal angles with the tangent line at that point. This is sometimes called the reflection property: if the ellipse were a mirror, a light ray emanating from the focus would be reflected to the other focus. (Then where would it go?)

Theorem 15(B). If a variable circle is tangent to a fixed circle and also passes through a fixed point outside the circle, then the locus of its moving center is a hyperbola.

In Fig. 54, O_1 is the fixed point, and

$$PO_2 - PO_1 = PO_2 - r = TO_2 = \text{constant}.$$

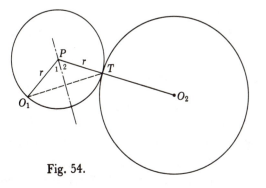

Fig. 54.

The corresponding paper model can be made by attaching to the circular disk a tab on which to place O_1. Then if a fold brings O_1 into coincidence with T, the crease passes through P and, by the argument used before, is tangent to the hyperbola at P. Since $\angle 1 = \angle 2$, the focal radii again make equal angles with the curve; but since they are on opposite sides of it there is no reflection this time.

As the point T is moved around the fixed circle, there comes a point beyond which it is no longer possible for the moving circle to be externally tangent at T. There are two such critical points, where tangent straight lines from O_1 touch the circle. That is, the moving circle becomes a straight line there, or it becomes a circle of infinite radius. Its center, which is supposed to lie always on the hyperbola, moves off to infinity. Thus the perpendicular bisectors of the tangent lines from O_1 are the asymptotes of the hyperbola. To "get on the other branch," P must be the center of a circle internally tangent to the fixed circle. I leave it to you to complete the interesting diagram.

The parabola is the easiest of all:

Theorem 15(C). If a variable circle is tangent to a fixed straight line and also passes through a fixed point not on the line, then the locus of its moving center is a parabola.

Proof (Fig. 55):

$$d_1 = r = d_2. \qquad \text{(Q.E.D.)}$$

Through considerations exactly like those for the ellipse, we obtain $\angle 1 = \angle 2$. This says that light originating at the focus is reflected by a parabolic mirror in a direction parallel to the axis of the parabola. Not only is this an interesting mathematical phenomenon; it also has immense practical

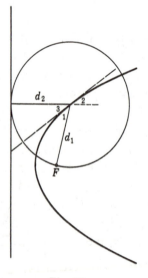

Fig. 55.

utility. It makes possible automobile headlights and search-
lights, whose reflectors are paraboloids of revolution, and it
is used in reverse by all parabolic telescopes and radar
screens.

CONFOCAL CONICS

From the reflection property we obtain

Theorem 16. (A) Any two intersecting confocal central
conics are orthogonal; (B) any two intersecting confocal
coaxial parabolas are orthogonal.

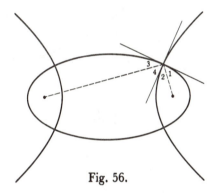

Fig. 56.

The "central" conics are ellipses and hyperbolas. Two
ellipses with the same foci never intersect each other; therefore one of the two confocal conics is an ellipse and the other
is a hyperbola, and we have a situation like that of Fig. 56.
The proof of Theorem 16(A) goes very quickly:

$$\angle 1 = \angle 3 \quad \text{(ellipse property)}$$
$$\angle 2 = \angle 4 \quad \text{(hyperbola property)}$$

Adding, $\angle 1 + \angle 2 = \angle 3 + \angle 4$

But two *equal* angles whose sum is a straight angle are both
right angles.

In part (B), by coaxial parabolas we mean two that share
the same axis of symmetry. If they are confocal they cannot
intersect unless they open in opposite directions, as in Fig. 57.

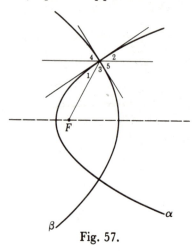

Fig. 57.

By the reflection property, applied to parabola α,
$$\angle 1 = \angle 2.$$
But with β as the reflector,
$$\angle 3 = \angle 4 = \angle 5.$$
Adding equals to equals,
$$\angle 1 + \angle 3 = \angle 2 + \angle 5$$
and each sum amounts to a right angle, as in part (A).

Because Theorem 16 holds for any two such conics, *every*
intersection in Fig. 58 is orthogonal.

PLANE SECTIONS OF A CONE

Three possible sections of a right circular cone by a plane are
shown in Fig. 59. A circle can be considered as a special
case·of Fig. 59A, where the cutting plane is perpendicular to
the axis of the cone. In Fig. 59B you must imagine that the

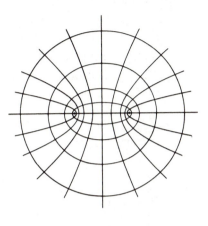
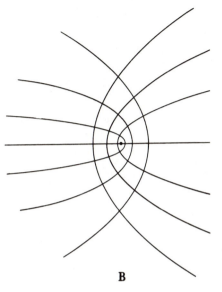

A   B

Fig. 58. Confocal conics.

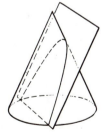
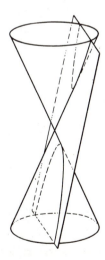

A   B   C

Fig. 59.

cone and plane, and hence the parabola, continue indefinitely downward. The plane is parallel to a slant element of the cone. In Fig. 59C, to get both branches of the hyperbola we need both *nappes* of the cone. Note that the plane need not be parallel to the axis of the cone. Again the figure must be extended indefinitely, this time both upward and downward.

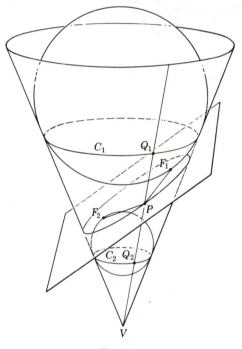

Fig. 60.

What do these pictures have to do with the locus definitions of the conics? We prove first that a section like that of Fig. 59A yields an ellipse.

In Fig. 60 there is just one sphere, called a Dandelin sphere after the inventor of this proof, that would fit snugly into the cone and also be tangent to the upper surface of the cutting plane, say at some point F_1. A second Dandelin sphere is tangent to the cone and to the plane at F_2. These two points will turn out to be the foci of the ellipse.

Select any point P on the curve and draw PV. Because P is on the cone, the straight line element PV not only cuts circles C_1 and C_2, but also is tangent to the two spheres, at Q_1 and Q_2. But PF_1 and PF_2 are also tangent lines to the spheres, because they lie in the tangent plane. Now tangents to any one sphere from an external point are equal. That is,

$$PF_1 = PQ_1$$
$$PF_2 = PQ_2$$

Adding, $\qquad PF_1 + PF_2 = PQ_1 + PQ_2 = Q_1Q_2.$

Now let P move around the ellipse. There will be a new Q_1 and a new Q_2 for each position of P, but the length Q_1Q_2 remains constant (why?). Thus

$$PF_1 + PF_2 = \text{constant},$$

the definition of an ellipse with foci at F_1 and F_2.

A beach ball rests on the floor, illuminated by a single electric light. What is the shape of the ball's shadow on the floor? Where does the ball touch the floor?

It is entirely possible to do the other two proofs with Dandelin spheres. The parabola has only one sphere (one focus), and for the hyperbola one sphere fits into each nappe. We leave the hyperbola proof for you to look up (see the Notes), or invent if you prefer. The parabola we will investigate by another method, devised in essence by Galileo, who lived two hundred years before Dandelin. First we have to do some (easy) analytic geometry.

You may remember that $y = x^2$, the simplest *quadratic function*, has a parabola for its graph. Fig. 61 gives the general idea. When $x = 0, y = 0$; when $x = 1, y = 1$; when $x = 2, y = 4$; and so on. A proof that the locus definition of a parabola is analytically equivalent to $ky = x^2$ is given in the Notes. In the parabola of Fig. 61, $k = 1$.

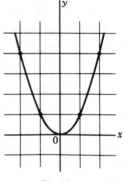

Fig. 61.

In Fig. 62 we seek the equation of the curve lying in a plane parallel to VB, an element of the cone. With origin at O, let P have coordinates (x, y) as indicated. $APBR$ is a circular section cut by a plane through P perpendicular to the axis of the cone. Take AB as the diameter perpendicular to PR. Then (cf. Fig. 4, page 11),

$$x^2 = AC \cdot CB.$$

Because $BCOQ$ is a parallelogram, $CB = OQ$. Making this substitution,

$$x^2 = AC \cdot OQ.$$

Also since OAC and VOQ are isosceles similar triangles,

$$\frac{AC}{OQ} = \frac{OC}{VQ} = \frac{y}{OV},$$

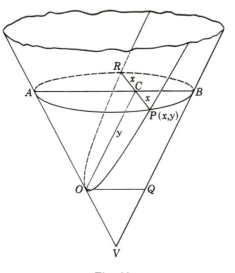

Fig. 62.

or
$$AC = \frac{OQ}{OV}y.$$

Putting this expression for AC into the last equation for x^2,

$$x^2 = \frac{OQ^2}{OV}y.$$

Which of these quantities are variables and which are
constants? We are talking about the equation of some one
curve (which we hope is a parabola). Therefore we do not
want to move the plane in which this curve lies nor change
the angle of the cone itself. If the horizontal circle is moved
up and down, the moving point P will have variable coordinates (x,y), which, however, will always be the coordinates of a point on the curve. Thus the constant OQ^2/OV
is what we called k in the equation $x^2 = ky$, and we know
that we have in fact a parabola.

A CHARACTERISTIC OF PARABOLAS

What is "similarity"? Two figures are similar when one can be made congruent to the other (or perhaps to a mirror image of the other) by a simple enlargement or contraction of the *scale*. Similarity can thus be considered "congruence except for size." For example, all circles are similar: they have the same *shape*.

It is evident (for instance, see Fig. 58A) that not all ellipses are similar. No simple magnification can turn a fat ellipse into a thin one, or vice versa. It would appear that the same can be said of parabolas—but this is not so. The wide parabolas of Fig. 58B *are exactly the same shape as the narrow ones*. This is a good illustration of an important mathematical rule: don't believe everything you see—or think you see. Many mathematicians would state it more strongly: don't believe *anything* you see. Of course, by this they mean: don't trust visual appearances. Everything must be proved.

The claim that a narrow parabola differs from a wide one only in size is readily substantiated. By changing the shape of the cone and the position of the cutting plane of Fig. 62 we can apparently obtain different parabolas. But our proof showed that $OQ^2/OV = k$ is the *only* quantity that will thus be altered. Now a change of scale (for instance in Fig. 61) multiplies x by some factor and multiplies y by the *same* factor. Let us submit the equation

$$x^2 = ky$$

to a change of scale by setting

$$x = kX$$
$$y = kY.$$

Making these substitutions,

$$(kX)^2 = k(kY)$$
$$k^2 X^2 = k^2 Y$$
$$X^2 = Y.$$

We have reduced our parabola to the curve of Fig. 55, except for size. But this can be done for any k. Hence all parabolas are simply variously scaled reproductions of Fig. 61.

7
Projective geometry

PROJECTIVE TRANSFORMATION

A projection from a point O in ordinary 3-space maps the points of a given plane into points of another given plane, provided O does not lie in either plane. In Fig. 63, triangle ABC of plane Σ (sigma) is projected to triangle $A'B'C'$ of plane Σ', O being called the center of projection. The two planes are not necessarily parallel, so in general the image of a figure is not similar to the original.

The transformation of Chapter 3, inversion, maps the plane onto itself. There is no reason why we can't use our principle of "Transform, solve, transform back again," when the transformation maps one plane onto another, so long as the transformation is one-to-one and reversible. One often speaks of "*the* projective plane," but the three-dimensional background is there none the less.

Considering our projective mapping as a transformation, we would like immediately to know what its invariants are. Already it is evident that whatever distortion takes place is different from that of the transformation of inversion in a circle. For one thing, straight lines are preserved, because any two points on a straight line in Σ, together with O, determine a plane, and this plane must intersect Σ' in another straight line. But angle is not invariant this time: the transformation is not conformal. This means that parallelism is not preserved, nor is similarity. But our persistent friend cross-ratio, which was preserved under inversion, is also a projective invariant, a significant circumstance to which we shall return.

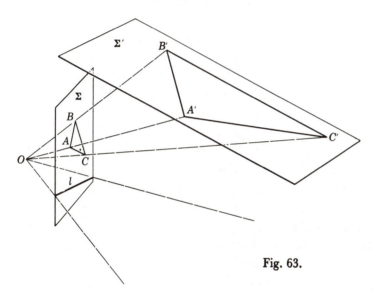

Fig. 63.

What about circles? They do not project into circles. But because they are sections of a cone, circles project into ellipses, hyperbolas, or parabolas. We saw this happening in the last chapter, where Σ was a plane perpendicular to the axis of the cone with vertex at O, and Σ' was the "cutting plane." We state without proof that it also happens if the cone is oblique to the circle. This has far-reaching consequences; among other things it means that any purely projective properties of circles automatically apply also to the conic sections.

What happens at infinity? Just as in the case of inversion, we must adopt a definition. Thus the question is not properly phrased. It should be, what do we *want* to have happen at infinity? It is up to us to make a consistent and workable definition that will fit the geometry without introducing contradictions or special cases. And it will not be the same definition that was useful for inversive geometry.

Note that under a projection (Fig. 63) each line of plane Σ projects to a line of plane Σ' *except one*. There is one straight line, l, that determines, together with O, a plane parallel to Σ'. What is the image of l on Σ'? We should prefer *every* line to project to a line; hence it is advantageous to stipulate that plane Σ' contains a line at infinity, the image of l. The correspondence is now complete and one-to-one between Σ and Σ'. It is worth observing that there is also one and only one line in Σ' that, with O, forms a plane parallel to Σ. This line projects into the line at infinity of the plane Σ. Thus each plane possesses one line at infinity.

Any two lines in a plane intersect in one and only one point. A side AB (extended) of triangle ABC in plane Σ intersects l in just one point. Therefore $A'B'$ must intersect l', the line at infinity of Σ', in just one point. No matter which of the two opposite directions you take along a given straight line, you will arrive at the same point on the line at infinity. If this seems paradoxical to you, it is only because you are still hampered by your preconceived notions of what infinity "is really like." We reiterate what we said in Chapter 3: infinity is not "really like" anything. We say with Humpty Dumpty, who has become so hackneyed in this regard that he must have fallen off that wall many times over in sheer despair, infinity "means just what I choose it to mean—neither more nor less." The ordinary Euclidean plane, together with its line at infinity, is called the *extended plane*, or simply the *projective plane*, and it has the great advantage of removing any special qualities from the line at infinity, which behaves for all projective purposes exactly like an ordinary line. For example, we can now speak of the unique point of intersection of *any* two (coplanar) lines. If the two lines are parallel in the Euclidean sense, their point of intersection lies on the line at infinity.

Before we explore further this new country called the projective plane, let us look at a theorem of projective geometry in 3-space suggested by Fig. 63.

Desargues' Two-triangle Theorem. If two triangles are in perspective from a point, the three intersections of their corresponding sides (extended) lie on one line.

The proof is easy. Planes Σ and Σ' must intersect in a line m, and that is the very line we want. For O, A, B, A', B' all lie in one plane. Therefore AB and $A'B'$ (extended) must intersect. These lines also lie in Σ and Σ', respectively, and hence their intersection occurs somewhere on m (the only line shared by Σ and Σ'). The same is true of the other two pairs of sides: all three pairs intersect on m.

The line m is in one respect the counterpart of the circle of inversion: every point on m is left invariant by the transformation.

Desargues' Theorem also holds in the plane, but interestingly enough its proof is more difficult in two dimensions than in three. To simplify it we shall resort to the now familiar device of "Transform, solve, transform back." Formerly we selected a suitable center of inversion. Now we seek a judicious center of projection and plane onto which to project.

Desargues' configuration in the plane could look something like Fig. 65, or it might look quite different; the projection can operate "both ways" through O, as in Fig. 64. (For instance, how are the points *below* l in plane Σ of Fig. 63 projected onto Σ'?) In either case, or a variety of others that you could draw, what we have to prove is that P, Q, and R are collinear.

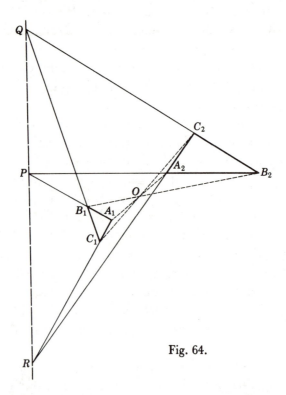

Fig. 64.

Looking once more at Fig. 63, we note that line l was sent to infinity by a projection from O upon Σ'. But *each* line of Σ, together with O, determines a new plane. Therefore by choosing a new Σ', parallel to such a plane, we can select a projectivity that sends any desired line to infinity.

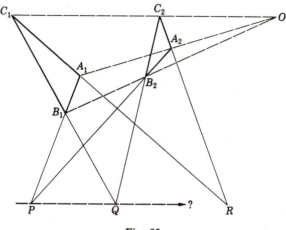

Fig. 65.

Returning now to Fig. 65, remember that *all* of its points lie in one plane, including the point O. Continuing the convention that primed capital letters stand for image points after projection, we have put no primed letters in Fig. 65. Using any center of projection not in this plane, *we select a projectivity that sends line PQ to infinity*. If we can prove that this projection also sends R to infinity then we know that P', Q', R' are collinear (all lying on the line at infinity), and hence that P, Q, R were collinear before projection.

After the projection P' lies on the line at infinity. But intersection is preserved, so that $A_1'B_1'$ and $A_2'B_2'$ must intersect at P'. The only way for two lines to intersect at infinity is for them to be parallel. The same applies to $B_1'C_1'$ and $B_2'C_2'$, because they intersect at Q'. Hence the transformed diagram looks like Fig. 66. Parallel lines cut off proportional intercepts on any two transversals. Hence,

$$\frac{s}{t} = \frac{x}{y}.$$

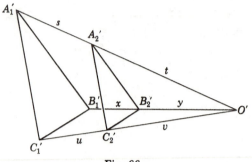

Fig. 66.

Likewise, $$\frac{u}{v} = \frac{x}{y}.$$

Therefore, $$\frac{s}{t} = \frac{u}{v},$$

which says that $A_1'C_1'$ is parallel to $A_2'C_2'$ and therefore that
their intersection R' lies on the line at infinity. This is what
we wanted to prove.

   Pappus' Theorem. If the vertices of a hexagon lie alternately

Fig. 67.

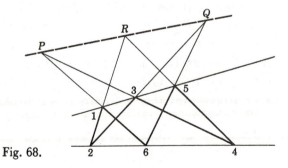

Fig. 68.

on two lines, then the three intersections of the pairs of opposite sides are collinear.

Pappus of Alexandria (300 A.D.) proved this theorem by laborious Euclidean methods. Perhaps you have already guessed how we can do it easily. The hexagon is re-entrant, intersecting itself (Fig. 68). To keep track of the vertices we number them from 1 to 6 in the order in which they appear when the hexagon is drawn without lifting the pencil. "Opposite" sides can then be picked out by associating the same numbering with the schematic model hexagon in Fig. 67.

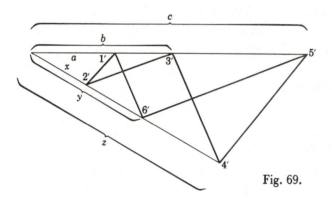

Fig. 69.

Just as in the Desargues proof, we project Fig. 68 so that *PQ* becomes the line at infinity. Then the pair of sides intersecting at *P* must become parallel and the pair intersecting at *Q* must become parallel, and we have the configuration of Fig. 69. Pappus required that the "two lines" of the theorem should not be parallel. This requirement is not necessary however; for if they were parallel before projection they (then) intersected on the (old) line at infinity of the plane. That line, after projection, must become an "ordinary" line, and the point of intersection an "ordinary" point on it.

Now in Fig. 69, because similar triangles are formed by
the parallels,

$$\frac{a}{b} = \frac{y}{z}, \qquad \text{or} \qquad a = \frac{by}{z}$$

Also

$$\frac{x}{y} = \frac{b}{c}, \qquad \text{or} \qquad x = \frac{yb}{c}$$

Dividing equals by equals,

$$\frac{a}{x} = \frac{c}{z}.$$

This tells us that lines 1'2' and 4'5' are parallel; that R' lies
on the line at infinity, collinear with P' and Q'; and hence
that P, Q, R were initially collinear.

Pappus' Theorem is interesting for its own sake, but it has
a much deeper significance.

THE FOUNDATIONS

For 2,000 years following the Greek era, mathematicians
were troubled by Euclid's parallel postulate, which states, in
effect, that through a point not on a given line one and only
one line can be drawn parallel to the given line. This is
more complicated than any of the other postulates and
axioms, and it was felt that somehow it should be possible
to derive this postulate from the others. Very much time and
effort were expended in the futile attempt to do so, until
finally, in the nineteenth century, several mathematicians
working independently took the opposite approach. They
thought: If it is indeed an independent postulate, and not a
consequence of the others, then why not toss it out from the
list and substitute something else in its place? They did so,

and were able to create whole new *non-Euclidean* geometries, containing no inner contradictions or inconsistencies. This took courage as well as genius, because Euclid's geometry had always been considered *the* geometry, inviolate and perfect. It seemed almost a sacrilege to tamper with it. It had never occurred to anyone that there could be other geometries. The new ideas were so bizarre that they were scarcely countenanced at first, and attracted little attention. Even the great Gauss was reluctant to publish his findings on the subject because of the furore he felt would be created.

When the significance of the new attack was finally grasped by other mathematicians, they began a thorough study of all the other fundamentals and foundations of geometry, and found, in places, rather shaky underpinnings. Several more postulates had to be added to make Euclidean geometry rigorous. Euclid thought of postulates and axioms as "self-evident truths." Today they are nothing of the sort. We have already seen how we can successfully postulate for one geometry a single point at infinity and for another geometry a straight line at infinity. These postulates are neither self-evident nor truths. They are, if you like, rules of the game.

The way the game is played today, and now I mean the whole of mathematics, is to begin by laying down the rules (axioms and postulates) and then to explore and exploit their consequences. The rules themselves are not mandated by anything: they can theoretically be any set of abstractions whatever—so long as they are consistent. What the mathematician does is try to select those and only those rules necessary and sufficient to do the job he wants done, to create the mathematics he is after. This is an oversimplified description of the axiomatic approach, but I hope not misleadingly so. "Playing the game" this way has resulted in clearing up a number of deeply concealed difficulties—and, to be sure, has created some new ones. Problems resulting from the study of the foundations of mathematics still draw the attention of many exceedingly talented people, and probably always will.

When this sort of overhauling of basic notions began, it soon became apparent that far too much had been taken for granted as "self-evident." The principal breakthrough of the non-Euclidean explorers occurred when they realized that *everything* had to be carefully scrutinized. Algebra also had to go onto the operating table and be dissected. For instance, $3 \times 5 = 5 \times 3$. The order of multiplication can be interchanged. This is called the *commutative* law of multiplication. "But of course," you are saying, "Why bother to give it a name? Multiplication has to be commutative." Ordinary multiplication of ordinary numbers, yes; but *all* multiplication? There are algebras and even geometries where multiplication is not commutative. Even the simplest examples are beyond our scope here without further background material; but for those who may have encountered *matrices*, we mention that matrix multiplication, leading to a host of useful and consistent applications, is not commutative. If (A) and (B) are matrices, $(A)(B) \neq (B)(A)$.

At the end of our proof of Pappus' Theorem we divided equals by equals to obtain

$$\frac{a}{x} = \frac{\dfrac{by}{z}}{\dfrac{yb}{c}} = \frac{by}{z} \cdot \frac{c}{yb} = \frac{c}{z}.$$

This was accomplished by canceling the *equal* quantities *by* and *yb*, which of course implies that multiplication is commutative. But what if it were not? Then the proof could not be completed, and Pappus' Theorem would not in fact hold.

We carefully displayed *by* and *yb* just as they occurred, one the reverse of the other. If you think you can prove Pappus' Theorem without invoking commutativity, it is because you have used it somewhere without noticing it. What we have, then, is not strictly a theorem but an *axiom* of projective geometry. We can add to whatever list of axioms we start with either an axiom of commutativity or Pappus' axiom: we must have one or the other.

Nothing akin to this discussion ever occurred to Pappus in Alexandria in 300 A.D.

CROSS-RATIO

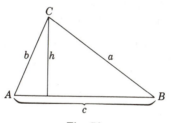

Fig. 70.

You may recall the "opposite over hypotenuse" definition of the sine of an angle. In Fig. 70,

$$\sin A = \frac{h}{b}, \qquad \text{so} \qquad h = b \sin A$$

$$\sin B = \frac{h}{a}, \qquad \text{so} \qquad h = a \sin B$$

Hence, $$b \sin A = a \sin B$$

or $$\frac{\sin A}{a} = \frac{\sin B}{b}$$

We can do a second pair in the same fashion, so that the complete Law of Sines, as it is called, says that in any triangle the sines of the angles are proportional to the opposite sides:

$$\frac{\sin A}{a} = \frac{\sin B}{b} = \frac{\sin C}{c}$$

Any four given points A, C, B, D on a line determine a cross-ratio,

$$\frac{AC}{BC} \Big/ \frac{AD}{BD}.$$

What we want to know is whether this ratio is preserved when the points are projected to another line from a point O (Fig. 71). We apply the Law of Sines a few times:

$$\frac{AC}{\sin 1} = \frac{OA}{\sin 5}, \qquad \therefore \qquad AC = \frac{OA \sin 1}{\sin 5}$$

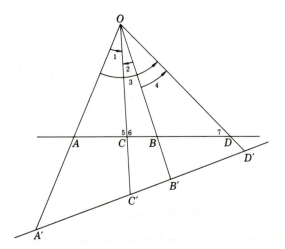

Fig. 71.

$$\frac{BC}{\sin 2} = \frac{OB}{\sin 6}, \qquad \therefore \quad BC = \frac{OB \sin 2}{\sin 6}$$

But the sines of *supplementary* angles, in this case 5 and 6, are equal. Therefore

$$\frac{AC}{BC} = \frac{OA \sin 1}{OB \sin 2}$$

Again,

$$\frac{AD}{\sin 3} = \frac{OA}{\sin 7}, \qquad \therefore \quad AD = \frac{OA \sin 3}{\sin 7}$$

$$\frac{BD}{\sin 4} = \frac{OB}{\sin 7}, \qquad \therefore \quad BD = \frac{OB \sin 4}{\sin 7}$$

Therefore

$$\frac{AD}{BD} = \frac{OA \sin 3}{OB \sin 4}$$

and finally,

$$\frac{AC}{BC} \div \frac{AD}{BD} = \frac{\sin 1}{\sin 2} \div \frac{\sin 3}{\sin 4}$$

The vital step is the last one: all reference to the transversal canceled out, meaning that the cross-ratio depends *only on the angles at O.* If we did the whole procedure using the primed points on the other transversal, it would all go exactly the same way and we should end with

$$\frac{A'C'}{B'C'} \div \frac{A'D'}{B'D'} = \frac{\sin 1}{\sin 2} \div \frac{\sin 3}{\sin 4}$$

which says that the cross-ratio is invariant under projection.

Note next that if we move O to a new location, P, without touching A, C, B, and D (Fig. 72), there will again be no

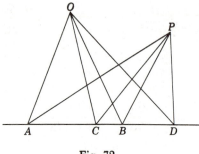

Fig. 72.

change whatever in the proof. Therefore, even though the individual angles are altered, the cross-ratio of their sines is not altered. For this reason we take this ratio as a *definition* of the cross-ratio of four concurrent lines.

We would like some abbreviations for the cross-ratio. For four collinear points, we shall write (AB, CD) in place of

$$\frac{AC}{BC} \bigg/ \frac{AD}{BD}.$$

The cross-ratio of the four lines concurrent at O we shall denote by $O(AB,CD)$. Then what we have been saying is summarized by

$$O(AB,CD) = (AB,CD)$$

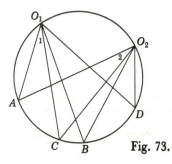

Fig. 73.

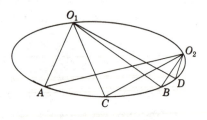

Fig. 74.

We can now talk about the cross-ratio of four points on a circle, which we postponed from Chapter 3. This is defined as the cross-ratio of four lines through the four points and concurrent in any point on the circle (Fig. 73). Because $\angle 1 = \angle 2$, etc. (measured by the same intercepted arc), the cross-ratio determined by the points A, C, B, and D is independent of the position of O, provided O stays on the circle. That is, $O_1(AB,CD) = O_2(AB,CD)$. Furthermore, we have established cross-ratio as a projective property; hence the same statement goes over by projection to a conic and applies equally to Fig. 74.

THE COMPLETE QUADRANGLE

We now recall that if the cross-ratio (AB,CD) is -1, the division is called harmonic. We observed in Chapter 3 that this is equivalent to the requirement that C and D be inverse points with respect to the circle whose diameter is AB. In particular, if C is the midpoint of AB, its inverse is the point at infinity. This checks with the original definition,

$$\frac{AC}{CB} = \frac{AD}{BD},$$

because as D recedes, AD and BD tend toward equality with each other (Fig. 75).

Fig. 75.

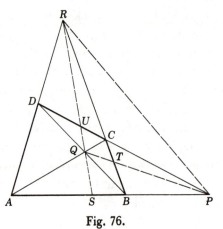

Fig. 76.

In a given quadrilateral *ABCD* (Fig. 76) there are three and only three ways of pairing the four vertices to form two intersecting lines:

> *AB* with *CD* to intersect in *P*
> *AC* with *BD* to intersect in *Q*
> *AD* with *BC* to intersect in *R*

Together with triangle *PQR*, the whole figure is called a *complete quadrangle*. It may seem rather remarkable that, no matter what the shape of the original quadrilateral, the

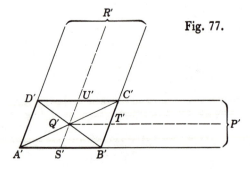

Fig. 77.

cross-ratios (AB,SP), and therefore also (DC,UP) by projection from R, and (BC,TR), and several others not shown lest the figure become too complicated, are all equal to -1. The reason is immediately evident if we project the line PR to infinity. Then the quadrilateral becomes a parallelogram (Fig. 77). Look at any one of the cross-ratios, say $(A'B',S'P')$. S' is now the midpoint of $A'B'$, because Q' is the intersection of the diagonals of a parallelogram; and P' is at infinity. Hence by the preceding paragraph, $(A'B',S'P') = -1$. But cross-ratio is invariant under projection; therefore $(AB,SP) = -1$.

A narrow straight irrigation ditch leads away from a storage tank in the middle of a large level field. The owner wants to dig a similar ditch on the other side of the tank, diametrically opposite, but the tank is too big to see over or around. He has no surveying equipment, but he has heard

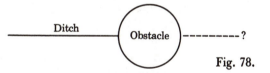

Fig. 78.

that you know something about geometry and he asks you to help him extend this straight line across the obstacle (Fig. 78). How can you solve this problem for him with nothing more than a bunch of marker pegs to drive into the ground? No string, no need for pencil and paper.

Put pegs at any two points A and B on the edge of the ditch, and select any two points C and D located approximately as shown in Fig. 79. Now find point R by visually aligning AD and BC. Q will be harder to find, requiring trial and error to get the lines straight because you don't have 180° simultaneous vision. (It's easy if you have two assistants to stand off at a distance on the extensions of BD

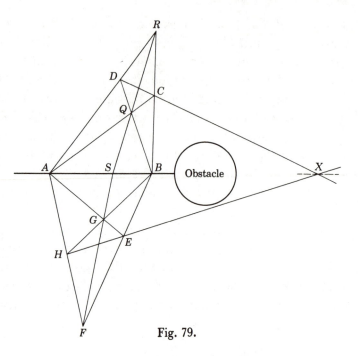

Fig. 79.

and *AC* and call instructions to you.) Point *Q* (with *R*) tells you where to put *S* on the edge of the ditch.

Next, step across to any suitable point and drive a stake for *E*. Put *F* in line with *BE*. Now find *G* as the crossing of the lines of sight *AE* and *SF*. *H* is the cross of *BG* with *AF*. And now you have *X*, by lining up *DC* and *HE* simultaneously.

Now the harmonic conjugate of *S* with respect to *AB* is some point *P* on the line you are trying to find. But *DC* passes through *P*. Likewise, so does *HE*. Therefore *P* = *X*, and your problem is half solved. Simply repeat the whole process with new *A*, *B*, *C*, and *D* to obtain a second point *Y*, and the straight line determined by *X* and *Y* is the edge of the extended ditch.

PASCAL'S THEOREM

Four fixed points on a line, in a given order, uniquely determine a cross-ratio. That is, if we are given three such points and a required cross-ratio, there is just one possible fourth point that can be used. This says that if two cross-ratios are known to be equal, and if three points of one coincide with three points of the other, then the fourth points coincide also. The same goes for cross-ratios of lines emanating from a common point. If, in Fig. 80, it is known that

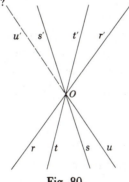

Fig. 80.

$O(rs,tu) = O(r's',t'u')$; and if it is known further that r and r' are parts of the same straight line, and that s and s' are parts of the same straight line, and that t and t' are parts of the same straight line; then it follows of necessity that u and u' also comprise parts of one straight line.

We shall now prove a generalization of Pappus' Theorem, known as

Pascal's Theorem. The points of intersection of the pairs of opposite sides of a hexagon inscribed in a conic are collinear.

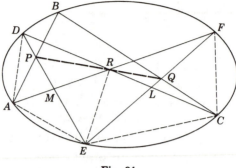

Fig. 81.

If the hexagon is *ABCDEF* in Fig. 81, then the points *P*, *Q*, and *R* are the ones that we wish to prove collinear. The other dotted lines of the figure are construction lines to aid in the proof. The hexagon need not be self-intersecting; if it is not, then the theorem of course implies that the pairs of opposite sides should be extended until they meet.

The theorem is proved by the following chain of equal cross-ratios:

$$R(FL,QE) = C(FL,QE) = C(FD,BE) = A(FD,BE) =$$
$$A(MD,PE) = R(MD,PE).$$

Now the first and last of these have three lines of the one coinciding with three lines of the other (cf. Fig. 80). Therefore the respective fourth lines coincide also, which says that *RQ* and *RP* are parts of the same straight line.

Why can't we prove Pascal's Theorem in the following obvious and tempting fashion? Start with a given conic and an inscribed hexagon whose opposite sides are *parallel*. The three intersections of these pairs lie on the line at infinity. Project that line back into an ordinary line. Then these intersections must remain collinear, the hexagon goes to another hexagon in a conic, and we seem to have proved the theorem. Well?

DUALITY

Let us say, for the moment, "two lines lie on a point" when we mean that the two lines have the point in common: they intersect at the point. Thus if three lines lie on a point, they are concurrent. Although it sounds awkward, this convention allows us to exhibit a *duality* between lines and points.

Two points determine a line.	Two lines determine a point.
Three points not lying on one line determine a triangle.	Three lines not lying on one point determine a triangle.
Four points lying on one line determine a certain cross-ratio. A fifth point not lying on that line determines, with the other four, four lines with the same cross-ratio.	Four lines lying on one point determine a certain cross-ratio. A fifth line not lying on that point determines, with the other four, four points with the same cross-ratio.

It will be observed that each statement in the second column is an exact replica of the corresponding statement in the first column, except for the interchange of the words "line" and "point" throughout. The third statement happens to be *self-dual*: one column repeats the information given in the other column.

Every purely projective theorem can be dualized. This very remarkable property cuts in half the number of proofs one needs. It is necessary to prove only one of the theorems of any pair: the other must automatically follow. Many a standard projective geometry textbook exhibits all the theorems dually in two columns, throughout the book. Projective duality has long been known to exist, but only in relatively recent times has it become thoroughly understood, as a consequence of the study of the foundations of mathematics. We build the whole geometry on the axioms. Therefore if we can set up all the axioms in such a way that they include their own duals, then every statement made in the geometry is automatically true in its dual form also. If the words *point* and *line* are interchangeable in the axioms, they are interchangeable everywhere.

Such a set of axioms, which will give us a large part of plane projective geometry, is the following:

1. One and only one line lies on every two distinct points, and one and only one point lies on every two distinct lines.
2. There exist two points and two lines such that each of the points lies on just one of the lines and each of the lines lies on just one of the points.
3. There exist two points and two lines, the points not lying on the lines, such that the point lying on the two lines lies on the line lying on the two points.

These axioms, or postulates if you prefer, do not seem to say very much; but they say just enough. An entire pure projective geometry can be built on them. (Note that they will not do for Euclidean plane geometry; Axiom 1 says that *every* two lines must intersect, even two parallel lines.)

That it is possible to write these axioms, and hence all the theorems derivable from them, in this dual form has a deep foundational significance. It means that projective statements about points and lines do not depend upon any intrinsic meaning expressed in the word "point" and the word "line." These words are *undefined* (Euclid ran into ludicrous difficulties when he tried to define them). We think we know, from long experience, what we mean when we say "point" and "line," but from the viewpoint of projective geometry we might just as well say "plop" and "lop," and the theorems would still be correct. We would no longer have the pretty pictures, because we couldn't draw a plop or a lop; but all the abstract statements about them would contain no contradictions or inconsistencies: the mathematical system would continue to flourish. This is comforting—and important—to know, because it means that none of the theorems depend on the diagrams.

It turns out that Desargues' and Pappus' Theorems have certain special metric (not purely projective) properties, as we have hinted, and cannot be derived from the above axioms. Nevertheless they both have valid duals. Desargues' Theorem dualizes to its converse (try it); but we have not previously encountered

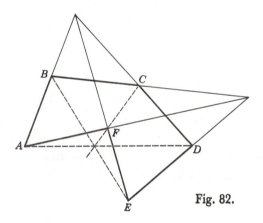

Fig. 82.

The Dual of Pappus' Theorem. If the sides of a hexagon are concurrent alternately in two points, then the three lines joining opposite vertices are concurrent (Fig. 82).

This suggests that a new theorem might be *discovered* by dualizing a known one. There is a famous example of just such a discovery. Two centuries after Pascal's time, C. J.

Brianchon, then a student of twenty at the École Poly-technique, dualized Pascal's Theorem to produce a new and very different one of his own:

Brianchon's Theorem. The lines joining the opposite vertices of a hexagon circumscribed about a conic are concurrent (Fig. 83).

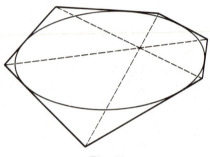

Fig. 83.

The dual of *points* lying on a conic is *lines* lying on a conic; that is, tangent to the conic. Thus the dual of an inscribed hexagon is a circumscribed hexagon. In the Notes we state the theorems in their dual forms.

As Pascal's Theorem generalizes Pappus' Theorem, so Brianchon's Theorem generalizes its dual. Indeed, the principle of duality tells us that this must happen.

8

Some Euclidean topics

The center of a circle tangent to two intersecting lines AB and AC must lie on the bisector of the angle at A (Fig. 84). If we add the requirement that it also be tangent to a third line BC then there are two such circles possible, with centers at I, called the *incenter* of triangle ABC, and E_a, one of its *excenters*.

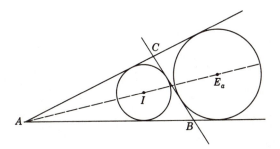

Fig. 84.

If we now repeat these statements with respect to vertex *B* we obtain the same incircle, plus the circle with center at E_b; and likewise for *C*, E_c. (Fig. 85). As we saw earlier in another connection, because $\angle 1 = \angle 2$ and $\angle 3 = \angle 5$, we have $\angle 3 = \angle 4$, and adding equals to equals,

$$\angle 1 + \angle 3 = \angle 2 + \angle 4$$

which means that $\angle 1 + \angle 3$ is a right angle. That is, E_aA is

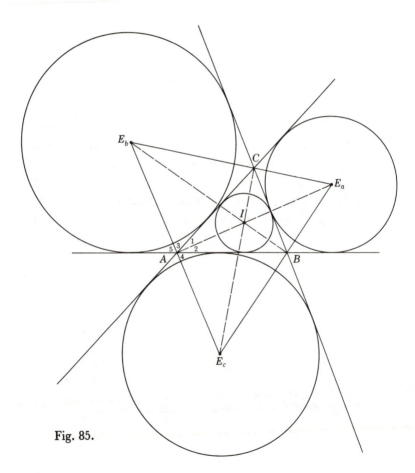

Fig. 85.

the altitude on E_bE_c; and we observe that, at least for this special triangle $E_aE_bE_c$, the altitudes are concurrent. We shall soon see that this happens in all triangles.

When a navigator at sea takes a star sight to plot the position of his ship, he measures with his sextant the angular distance of the star above the horizon. There is a line on the surface of the ocean (actually a circle on the earth's surface) that is the locus of points from which this angle can be observed at any given time. With the aid of navigation tables he plots this line on his chart. Theoretically, the ship should lie somewhere on it. He repeats the process with another star, and where the two lines cross is his position, or "fix."

In practice he usually takes three sights and plots all three lines. Because of instrumental limitations, inaccuracy of chronometers, an unstable platform (the ship), and other factors, not the least of which is the possibility of human error, he seldom obtains a perfect fix. Instead of intersecting, the three lines form a triangle, which he hopes will be reasonably small. If it is, he places the ship in position I in the triangle (equidistant from all three lines), and considers that he has a good fix.

It is interesting to note that although I is often the ship's most likely position, it may not always be. Star sights with a marine sextant must be taken during a span of only a few minutes at twilight (or dawn) when there is just enough light to make the horizon visible. Let us suppose that the southern half of the sky is cloud covered, but that three good stars are available, all within a sector roughly northwest to northeast. From these the navigator obtains the three lines of position AB, BC, and CA of Fig. 86. Before he jumps to the conclusion that the ship is probably at I, he should at least consider the possibility of E_c. If he chooses I, he assumes that his sights show three random errors, one away from the star and two

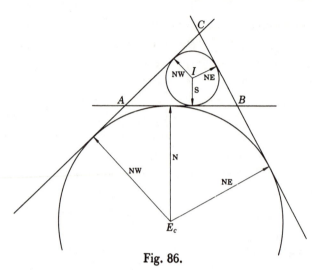

Fig. 86.

toward the star. But it is also possible to have systematic errors, from such causes as an undetected misalignment of the sextant, a tendency to sight too high (or too low), or a uniform misreading of the instrument. This last, possible with the vernier of an older sextant, is admittedly much less likely with the modern micrometer. Such a systematic error could make E_c the more likely fix because the displacements are all in the same direction (towards the star). Inasmuch as I and E_c can be several miles apart, it is worth an effort to avoid this disquieting possibility. The best way out is to choose three stars not in the same sector of the sky. If line AB had been plotted from a star observed *due south* instead of north, then the point I would be the most likely position even if there were a systematic error.

Small boats at sea today, particularly in foggy weather, rely heavily on RDF (Radio Direction Finder) for their fixes. The RDF instrument is quite prone to a uniform error, either built in or due to the heavy hand of an inexperienced

operator, always to the right or always to the left. Furthermore, the only three stations within range of your instrument may be (usually are for American yachtsmen) on one
shore, all within a single 180° sector. This is a natural for the
excircle. The three lines of position are now the supposed
directions of stations X, Y, and Z (Fig. 87). Position E_c
should be seriously considered; it is the only point equidistant from *and on the same side of* all three bearing lines.

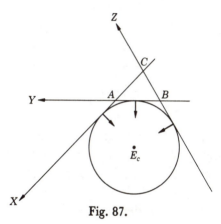

Fig. 87.

A THREE-CIRCLE PROBLEM

Prove that the common external tangents to each pair of
three different-sized circles meet in three collinear points.

In Fig. 88, P, Q, R are to be proven collinear. There is a
short and simple proof, which is not easy to invent from
scratch, however, because it depends on the introduction of
an auxiliary circle. With this hint, perhaps you should try
to do it before turning to the Notes.

We had a powerful projective method for proving col-
linearity, but it won't work here. If we project line PQ to

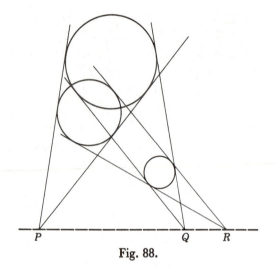

Fig. 88.

$P'Q'$ at infinity, the pair of tangents intersecting at P' are parallel; likewise the pair intersecting at Q' are parallel. This would seem to be saying that the three circles have become circles of equal diameter. Therefore the tangents that intersect at R' are also parallel, which puts R' on the line at infinity along with P' and Q', and hence PQR was initially a straight line. The flaw lies in forgetting that circles do not in general project to circles, but to other conics. Hence it does not follow that R' is at infinity after the projection.

The most striking way to prove the theorem is through its three-dimensional counterpart: The enveloping tangent cones of each pair of three different-sized spheres have collinear vertices. To demonstrate this, think of the three spheres as lying on a plane (a table top, for instance). Then there is just one other plane tangent to the three spheres in such a way that all three lie on the same side of it ("underneath" it). The three cones are also tangent to both planes, which intersect in a line. Hence the vertices of the three cones lie on that line also. Now the plane through the centers of the three spheres is the plane in which Fig. 88 lies.

THE EULER LINE

If *M* is the midpoint of *BC* (Fig. 89), then *AM* is called a *median* of the triangle. Let *BN* be a second median. Let *Q* and *R* be the midpoints of *AG* and *BG*, respectively, and draw *QR*. Now *MN* is a line parallel to side *AB* and bisecting the

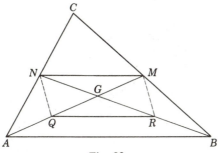

Fig. 89.

other two sides of triangle *ABC*; hence it is half as long as *AB*. *QR* does the same in triangle *ABG*. Hence *MN* and *QR* are equal and parallel, making *MNQR* a parallelogram. But the diagonals of a parallelogram bisect each other, so *AQ* = *QG* = *GM*. Thus *G* is a point of trisection for each of the two medians *AM* and *BN*, and, in like manner, *G* is a point of trisection for the third median.

G is called the *centroid*, which to the physicist means the center of gravity of the triangle considered as a thin plate of uniform density.

The *circumcenter P* is the center of the circle circumscribed around a triangle. There is only one *circumcircle*, because three points determine a circle. *P* is thus the only point equidistant from the three vertices, and hence (why?) is the intersection of the perpendicular bisectors of the three sides.

In the course of the next proof it will be shown that the three altitudes are always concurrent, in a point called the *orthocenter, O.*

In any (except an equilateral) triangle, find P and G, and extend the segment PG to the point O, where $OG = 2GP$ (Fig. 90). Now $CG = 2GM$, by the centroid property, and

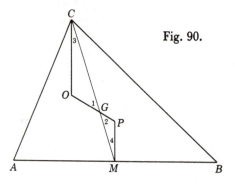

Fig. 90.

$\angle 1 = \angle 2$. Therefore triangles OGC and PGM are similar, making $\angle 3 = \angle 4$. These are alternate interior angles and therefore CO is parallel to PM and hence perpendicular to AB. Thus CO is (part of) an altitude. But O was uniquely determined by P and G; hence the proof can be repeated with respect to AO and BO, and we have

(1) The three altitudes are concurrent at O.

(2) The orthocenter O, the centroid G, and the circumcenter P are collinear. OGP is called the *Euler line.*

(3) G is a trisection point of OP.

Inasmuch as (2) states that three points lie on a line, you might be tempted to try dualizing it to discover some three lines that pass through a point. Do you see at once why this is impossible? The proof is chock-full of non-projective ideas:

perpendicularity, lengths two-thirds of other lengths, similarity of triangles, equality of angles—all properties not
preserved under projection. This proof points up some of the
differences between Euclidean geometry and projective
geometry.

THE NINE-POINT CIRCLE

In any triangle ABC (Fig. 91), the midpoints of the three
sides (A',B',C'), the midpoints of the three lines joining the
orthocenter to the vertices (A'',B'',C''), and the feet of the
three altitudes (D,E,F), all lie on one circle.

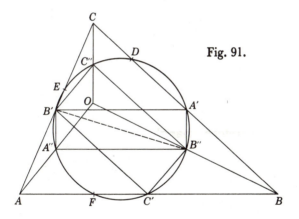

Fig. 91.

Proof. For the same reason as in Fig. 89, we have a parallelogram $A'B'A''B''$; but this time two sides are parallel to
CO which is perpendicular to the other two sides, so that
$A'B'A''B''$ is a rectangle. Exactly similarly, $B'C'B''C''$ is a
rectangle. The two rectangles have a common diagonal,
$B'B''$. By the corollary of Theorem 1, (page 7), this is
enough to put A', A'', C', and C'' all on the circle whose
diameter is $B'B''$. Furthermore, $C'C''$ is another diameter of

that circle; $C''FC'$ is a right angle; therefore F lies on the circle. In like manner, $B'EB''$ and $A'DA''$ take care of E and D.

The nine-point circle is the circumcircle of triangle $A'B'C'$, which is similar to triangle ABC with sides half the length of those of ABC, respectively. Thus the nine-point circle has radius equal to one-half the radius of the circumcircle of the original triangle (Fig. 92).

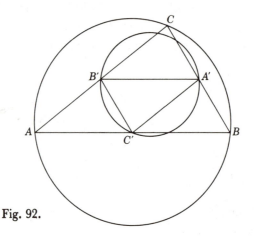

Fig. 92.

A TRIANGLE PROBLEM

The following theorem was proposed as a problem in the *American Mathematical Monthly* in 1968. Starting with any triangle ABC, construct the (exterior) squares $BCDE$, $ACFG$, and $BAHK$; then construct parallelograms $FCDQ$ and $EBKP$. The problem is to prove that PAQ is an isosceles right triangle.

When you get the drawing done it looks something like a Pythagorean theorem diagram because of the squares on the three sides. But *ABC* is not a right triangle, it is just *any* triangle. It seems hardly credible that something as metrically precise as an isosceles right triangle would emerge. But it always does, and the proof can be accomplished using only the most elementary high school geometry. You might try it as a (not easy) test of your ingenuity before reading further. . . .

~~Any luck?~~ If not, here is a large hint: draw diagonals *BP* and *CQ* of the parallelograms.

In this chapter we have touched on a very few topics of what might be called advanced classical school geometry. The subject has been widely developed. The theorems about triangles and circles continue almost without end, extension building on extension, as geometers find more and more relationships among special points, lines, triangles, and circles—and then planes, tetrahedra, and spheres when another dimension is added. For those who take kindly to this type of mathematics and would like to see some more of it, we list further references in the Notes.

9
The golden section

If we inscribe a regular pentagon in a circle and draw all its diagonals, we have (Fig. 93) so many interconnecting relations among the line segments that the diagram has earned the name of ~~mystic pentagram~~. Triangles *ACD* and *CDQ* are

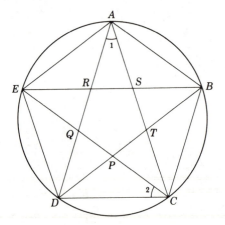

Fig. 93.

similar, because $\angle 1 = \angle 2$ (why?) and they share the angle at D. Further, they are both isosceles, because $AD = AC$. Proportional sides yield

$$\frac{AD}{QC} = \frac{QC}{QD}.$$

But $QC = AQ$ (why?), and therefore, substituting,

$$\frac{AD}{AQ} = \frac{AQ}{QD}. \tag{1}$$

Thus diagonal AD is divided in "extreme and mean ratio": the longer segment is the mean proportional between the whole diagonal and the shorter segment. The Greeks called such a division a Golden Section.

If we let $AD/AQ = \varphi$, then if the side of the pentagon is taken as unity, $AQ = 1$, $AD = \varphi$, and $QD = \varphi - 1$. Then, from equation (1),

$$\varphi = \frac{1}{\varphi - 1} \tag{2}$$

or, $\varphi^2 - \varphi - 1 = 0.$

We are interested in the positive root of this quadratic equation,

$$\varphi = \frac{1 + \sqrt{5}}{2} = 1.62 \text{ (approximately)}.$$

Triangle ADC (*et al.*) is isosceles with vertex angle half the size of its base angles (compare intercepted arcs). But

$$A + 2A + 2A = 180°,$$

the sum of the angles of any triangle. That is, $5A = 180°$, $A = 36°$. Therefore, in order to inscribe a regular pentagon in a given circle, we want a ruler and compass construction for an angle of 36°. If we draw a right triangle of base 1 and altitude $\frac{1}{2}$ (Fig. 94), then the hypotenuse is $\sqrt{5}/2$. Then

$$a = \frac{\sqrt{5}}{2} - \frac{1}{2},$$

and

$$b = 1-a = \frac{3}{2} - \frac{\sqrt{5}}{2}.$$

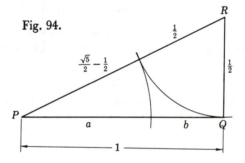

Fig. 94.

I leave it to you to carry through the arithmetical check that $1/a = a/b$. That is, the unit line PQ has been divided in extreme and mean ratio. It only remains to construct an isosceles triangle with unit equal sides and base a to obtain the 36° angle. Then *any* 36° angle inscribed in a circle will subtend an arc whose chord is a side of the pentagon.

If you take a long ribbon of paper and tie a simple over-hand knot in it, you will find, if you pull it taut carefully, that it flattens into a regular pentagon (Fig. 95). Why do you think this happens? "Probably because the knot requires five folds." So we count them and find only three. "Then there must be five diagonals." No: only two diagonals are terminated by folds, two more play an ancillary role, and the fifth one doesn't appear at all. Is it only pure coincidence that creates this perfect little pentagon? Is there another kind of knot that produces a polygon of six sides? Seven? I do not know the answers to these questions. The mathematical theory of knots is a new and difficult field, and whether this is part of that theory or only an insignificant happenstance is not yet known.

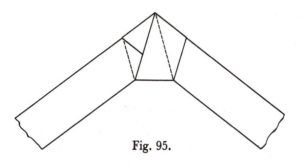

Fig. 95.

SIMILARITIES AND SPIRALS

If we draw a rectangle of base φ and altitude 1, and cut from it a unit square as shown in Fig. 96, then the defining relation for φ (equation 2) gives

$$\frac{\varphi}{1} = \frac{1}{\varphi - 1}$$

That is, the rectangle remaining also has its sides in the ratio of φ:1. Therefore the process can be repeated, the fourth new rectangle being similarly oriented to the original one. The continued similarities require that diagonals AD and BE serve also as diagonals of all the rectangles, and hence that their intersection O is the limit point around which all are nested.

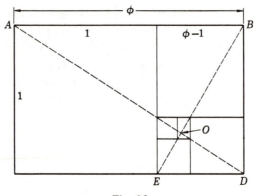

Fig. 96.

An analogous procedure yields the nested triangles of Fig. 97. Here the similarity occurs after a rotation of 108° instead of 90°. Triangle *FGH* is the fifth in the sequence (not counting *ABC* itself), so that it is not until we reach the tenth triangle that we find the orientation to be the same as that of *ABC*. The intersection of any two of the ten possible lines connecting a vertex with that of the next oppositely oriented triangle, like *AF* and *BG*, intersect in the limit point *O* which is common to all.

Selecting *OB* as a unit along a polar axis, we can find the equation in polar coordinates of a single continuous spiral curve passing through *A, B, C, D,* ... indefinitely. If θ is the positive trigonometric angle at *O*, then *r*, the distance from *O*, increases by a factor of φ for every 108° turn and decreases by the same factor for every −108° turn. Since 108° = 3π/5 radians, the desired equation of the spiral is

$$r = \varphi^{(5/3\pi)\theta}.$$

If θ goes through two magnification units, that is, θ = 2(3π/5), then *r* = φ², etc.

In Fig. 96, similarity occurs every $90° = \pi/2$, so that the equation of its spiral would be

$$r = \varphi^{(2/\pi)\theta}$$

The slightly larger exponent means that the coils are more widely spaced than those of the first spiral when the two are scaled to the same units.

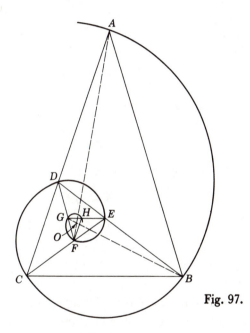

Fig. 97.

Because of the five-way symmetry there are various spirals to choose from in the pentagonal case. For example, in Fig. 93 the similarity between $ABCDE$ and $PQRST$ (in that order) depends on a rotation of $180°$, or π radians. But the similarity factor is no longer φ but

$$\frac{\varphi}{\varphi-1} = \varphi \cdot \frac{1}{\varphi-1} = \varphi^2, \quad \text{by (2).}$$

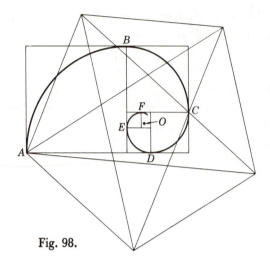

Fig. 98.

Therefore the spiral's equation is

$$r = \left(\varphi^2\right)^{(1/\pi)\theta} = \varphi^{(2/\pi)\theta}$$

That this is the same as that obtained for the rectangle explains why Fig. 98 is possible. Point C is a vertex of the interior pentagon opposite to A, and E would fall on the vertex of the next smaller pentagon (not drawn) opposite to C. The limit point O of the rectangles and the pentagons is the center of the circle circumscribing any pentagon.

These exponential spirals are selected by nature for the Chambered Nautilus and other sea shells, as well as for the more humble garden snail. It would be unnatural for the animal to build his portable house in a spiral that lacked the similarity property. What is required is a structure that, while growing with the animal, will maintain the same shape. The exponential spiral is the only kind that accurately answers this demand.

THE REGULAR POLYHEDRA

A regular convex polyhedron has congruent regular polygons for its faces and congruent multihedral angles at its vertices. There are only five such polyhedra, which we tabulate according to their numbers of faces and vertices:

POLYHEDRON	FACES	VERTICES	KIND OF FACE
Tetrahedron	4	4	Triangle
Hexahedron (Cube)	6	8	Square
Octahedron	8	6	Triangle
Dodecahedron	12	20	Pentagon
Icosahedron	20	12	Triangle

Why is it so certain that there are only five? There are an *infinite* number of regular polygons; why no more polyhedra? The formation of corners, together with the last column of the table, gives the answer. All the face angles at any one vertex must total *less* than 360°. For example, it is not possible to have another solid, in addition to the cube, bounded by squares. Four squares cornering together total 360°, and such a vertex would "flatten out"; this is the limiting case,

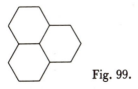

Fig. 99.

and it does not form a vertex of a polyhedron. Three hexagons can be grouped together (Fig. 99), but again the entire 360° are used up and there is no way for the vertex to be a meeting point of three non-coplanar faces. The situation for three polygons of seven or more sides is worse: they cannot even be grouped around a single point in a plane. This leaves us only pentagons, squares, and triangles to work with. Three pentagons can form a vertex, but four are too many. Thus we have just one possible regular polyhedron with pentagonal faces, and, for the same reason, only one with square faces. But three, four, or five equilateral triangles can be used at a single vertex, yielding three different solids with triangular faces. That completes the collection.

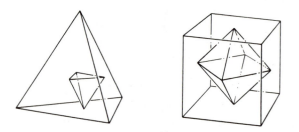

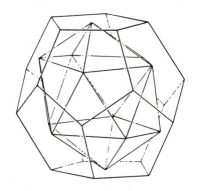

Fig. 100.

These solids are certainly objects of metric, as opposed to projective, geometry. Yet they exhibit a kind of duality, which you may already have observed in the table. Each is the dual of another with respect to the interchange of number of faces and number of vertices. If the tetrahedron is allowed to count twice, as its own dual, there are three dual pairs. They are arranged in Fig. 100 to exhibit the duality. Each vertex of the inner polyhedron is the incenter of a face of the outer polyhedron.

Fig. 101.

During the fifteenth century Fra Luca Paccioli, in a book titled *De Divina Proportione*, described thirteen different *effects* (properties) of the Golden Ratio φ. The twelfth "almost incomprehensible effect" involves the regular icosahedron. The five triangular faces at any vertex form a pyramid based on a regular pentagon. Any two opposite edges of the icosahedron are the sides of a rectangle whose longer sides are diagonals of one of the pentagons. But from our discussion of the pentagram we know that such a rectangle is a golden one. It has for its vertices four vertices of the icosahedron. The remaining eight must be symmetrically located vertices of two more such rectangles, which means that the twelve vertices of the three mutually orthogonal golden rectangles of Fig. 101 are the vertices of a regular icosahedron. By the duality property, of course, they are likewise the incenters of the faces of a dodecahedron.

THE CONTINUED FRACTION FOR φ

From the defining relation for φ we have

$$\varphi^2 - \varphi - 1 = 0$$

or $$\varphi^2 = \varphi + 1.$$

Dividing through by φ,

$$\varphi = 1 + \frac{1}{\varphi}$$

Now make what the computer people might call a loop: put the expression for φ into itself. That is, substitute the right-hand side *into* the right-hand side:

$$\varphi = 1 + \frac{1}{\varphi} = 1 + \cfrac{1}{1 + \cfrac{1}{\varphi}}$$

Keep repeating this process indefinitely to obtain the *continued fraction* for φ:

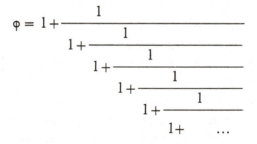

$$\varphi = 1 + \cfrac{1}{1 + \cfrac{1}{1 + \cfrac{1}{1 + \cfrac{1}{1 + \cfrac{1}{1 + \cdots}}}}}$$

To say that this expression converges to φ means that we can write a *finite* (terminating) continued fraction whose value is as close as we please to that of φ itself, simply by giving the finite fraction a sufficiently high number of "levels." Convergence can be proven, but we won't attempt it here. We look only for a moment at the first few *convergents*, or partial continued fractions:

$$1 = \frac{1}{1} = 1.0$$

$$1+\frac{1}{1} = \frac{2}{1} = 2.0$$

$$1+\cfrac{1}{1+\cfrac{1}{1}} = \frac{3}{2} = 1.5$$

$$1+\cfrac{1}{1+\cfrac{1}{1+\cfrac{1}{1}}} = \frac{5}{3} = 1.67$$

It should be evident that we are doing this the hard way. Each convergent c_n is expressible in terms of the previous one, c_{n-1}:

$$c_n = 1+ \frac{1}{c_{n-1}}$$

The sequence is much more easily extended in this way:

$$1+\frac{2}{3} = \frac{5}{3} = 1.67$$

$$1+\frac{3}{5} = \frac{8}{5} = 1.6$$

$$1+\frac{5}{8} = \frac{13}{8} = 1.625$$

$$1+\frac{8}{13} = \frac{21}{13} = 1.615$$

These numbers are approaching φ, closing down on φ alternately from above and below. This is all part of a longer story. But it is easy to see the law of formation of the sequence of numerators (*and* the sequence of denominators) of the fractions: each is obtained by adding the two previous ones. They are the famous Fibonacci Numbers,

$$1, 1, 2, 3, 5, 8, 13, 21, 34, 55, \ldots$$

Any further pursuit of this subject would lead us too far away from geometry. We mention all these things only because they exhibit a striking connection across parts of algebra, number theory, geometry, and analysis, the four major branches of mathematics that were once thought to be essentially distinct from one another. During the twentieth century it has become increasingly apparent that almost no topic is completely isolated in mathematics; and the interest and importance of a problem are greatly enhanced by its significant interconnections.

1 0

Angle trisection

Although we often read of "the three" great unsolved problems of antiquity, there were really four: squaring the circle, duplication of the cube, trisection of an angle, and the construction of the general regular polygon. All were to be solved with straightedge and compass, the only permitted tools of Euclidean geometry. For two thousand years they remained unsolved, but during the nineteenth century all were proved unsolvable. To prove that a problem *cannot* be solved under the prescribed conditions and rules of the game is in itself equivalent to a solution, in that the problem is thus removed from the docket of unfinished business once and for all.

The reason it took so long to dispose of these four problems is that they were being attacked in the wrong way. It is likely that their solvability could never have been determined solely within the confines of synthetic (Euclidean) geometry. It was not until the problems were embedded in the much more flexible framework of analysis that it was possible to

135

remove the obscuring cloak of ruler and compass and see what kinds of things were really causing the difficulties. The necessary analysis is slightly beyond our scope here; therefore we shall outline only in general terms the methods of attack, referring you to other sources for the details.

The most intractable of the four problems turned out to be the first. "Squaring the circle" has become the popular name for the process of finding the side of a square whose area shall be equal to the area of a given circle. Being free to choose our scale, we let the radius of the circle be unity. Then its area (πr^2) is π, and what we seek is a ruler and compass construction for $\sqrt{\pi}$, or for π itself.

The solutions to all of these problems remained tightly locked up until mathematics progressed to the point where the problems could be stated in algebraic instead of geometric form. First it was necessary to discover what constructions were possible with ruler and compass in terms of coordinate (analytic) geometry. One soon finds, without any very deep investigations, that all such constructions can be represented by rational operations and combinations of *square* roots, but that no such construction can produce any *cube* roots. If, then, any one of the four problems involves finding a cube root (or something worse), and if this involvement can be shown to be unavoidable, then the problem is forever unsolvable with ruler and compass.

Combinations of the various nth roots, and, in particular, the constructible combinations of square roots, are always solutions to a certain definitely specifiable type of equation called algebraic. In 1882, after much hard labor, it was finally proved that π is not the solution of any algebraic equation; thus it is not a number constructible with ruler and compass.

To duplicate the cube means to find the edge of a cube

whose volume is twice that of a given cube. If the given cube has unit edge, and hence unit volume, then the required cube, of volume 2, has edge $\sqrt[3]{2}$. But we know that cube roots cannot be constructed with ruler and compass.

The trisection proof contains trigonometry as well as algebra. An easily derived trigonometric identity states that

$$4(\cos \theta/3)^3 - 3(\cos \theta/3) = \cos \theta$$

Suppose we are asked to trisect an angle of 60°, whose cosine is $\frac{1}{2}$. That means we must find x, where

$$4x^3 - 3x = \tfrac{1}{2}$$

We then have to show that this cubic equation has no roots that are rational numbers or rational combinations of square roots. Then an angle of 60° is not trisectable with ruler and compass. But if there were any *general* process for trisection with ruler and compass, it would have to include the 60° angle. This proves that there is no such general process.

The fourth problem asks what regular polygons are constructible with ruler and compass. The Greeks knew how to construct the regular n-gon with $n = 3$, 4, 5, or 6. By inscribing these polygons in circles and repeatedly bisecting the sides, one gets polygons of $n \cdot 2^m$ sides, where m is the number of bisections. But what about the regular polygon of 7 sides, or 9, or 11? The problem was first solved by Gauss, who proved that the constructibility was intimately connected with the Fermat numbers.

Pierre de Fermat, who died a century before Gauss was born, made a conjecture concerning numbers of the form $2^{2^n} + 1$.

$$F_0 = 2^1 + 1 = 3$$
$$F_1 = 2^2 + 1 = 5$$
$$F_2 = 2^4 + 1 = 17$$

$$F_3 = 2^8 + 1 = 257$$
$$F_4 = 2^{16} + 1 = 65537$$
$$F_5 = 2^{32} + 1 = 4294967297$$

Fermat knew that all of these up to and including F_4 were prime numbers. He *guessed* that *all* F_n were prime; but this guess was incorrect. F_5 has the factors, discovered by Euler, 641 and 6700417, and to date no further prime Fermat numbers have been located. It is quite possible that all the rest are composite.

Gauss's remarkable achievement was to show that it is possible to divide the circumference of a circle into n equal parts when n is odd, if n is either a prime Fermat number or a product of different prime Fermat numbers. This he accomplished at the age of eighteen, and if he had ever previously doubted it, he knew then that mathematics was to be his life.

The regular 7-gon is not constructible with ruler and compass, nor is the regular polygon of 9 sides, nor 11. The next possible (odd) one is 15 (product of 3 and 5), and then 17. The non-constructibility of the regular 9-gon incidentally proves the impossibility of angle trisection. For if it were possible to trisect an angle of 60°, one could obtain an angle of 40°. But 40° is the central angle subtended by the side of an inscribed regular 9-gon, and the 9-gon would then be constructible.

OTHER KINDS OF TRISECTIONS

Although one cannot trisect an angle with ruler and compass, there are many other well-known devices for doing so.

1. (Archimedes) Inscribe the given angle θ in a circle (Fig. 102). Place marks on a straightedge so that the distance

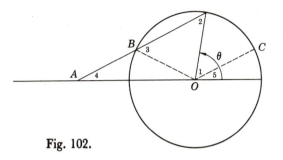

Fig. 102.

AB between the marks is equal to the radius of the circle. Then find the position (there is only one) such that the straightedge lies as in Fig. 102. Then the angle at *A* is ⅓θ. Of course marked rulers that are positioned by trial and error are taboo in Euclidean geometry.

Proof. Draw *BO*, and draw *OC* parallel to *AB* Then

$$\angle 1 = \angle 2 = \angle 3 = 2\angle 4 = 2\angle 5.$$

2. ("The Tomahawk") Attach a semicircular disc to a T-square, with dimensions as shown in Fig. 103. This is a trisector when placed on an angle as in Fig. 104.

Proof. Triangles *OAB*, *OCB*, and *OCD* are all congruent. Hence

$$\angle 1 = \angle 2 = \angle 3.$$

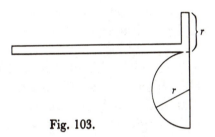

Fig. 103.

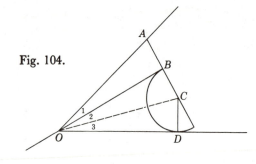

Fig. 104.

3. (Nichomedes) Through any point P on side OP of $\angle AOP$ (Fig. 105), draw two lines, one perpendicular to OA and one parallel to OA. Then place OB so that $BC = 2OP$. $\angle 4$ now trisects $\angle AOP$.

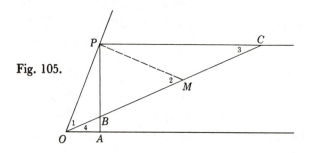

Fig. 105.

Proof. Connect P with M, the midpoint of BC. Then

$$\angle 1 = \angle 2 = 2\angle 3 = 2\angle 4.$$

Mathematically this method is the same as that of Archimedes (1). Turn Fig. 105 upside down and compare it with Fig. 102.

4. (Successive bisection) Bisect angle AOB (Fig. 106) with line (1). Then bisect the *lower* half angle with line (2); then the *upper* half of the angle last bisected, with (3); then the *lower* half of the angle last bisected, with (4); and so on

with alternating upper and lower bisections. The lines (n) approach the trisector of angle AOB as n increases. In a sense this is the best method, for it allows us to get *arbitrarily close* to the trisection with ruler and compass properly used.

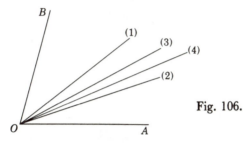

Fig. 106.

Proof. If $OAB = \theta$, we have after the first bisection an angle of $\frac{\theta}{2}$. From this we subtract $\frac{1}{2}(\frac{\theta}{2})$, then add $\frac{1}{2}(\frac{1}{2}(\frac{\theta}{2}))$, then subtract $\frac{1}{2}(\frac{1}{2}(\frac{1}{2}(\frac{\theta}{2})))$, and so on. After n bisections we have an angle equal to

$$\alpha_n = \frac{\theta}{2} - \frac{\theta}{4} + \frac{\theta}{8} - \frac{\theta}{16} + \cdots \pm \frac{\theta}{2^n}$$

$$= \theta \left[\frac{1}{2} - \frac{1}{4} + \frac{1}{8} - \frac{1}{16} + \cdots \pm \frac{1}{2^n} \right]$$

But the limit of the series in the brackets, as $n \to \infty$, is, by the geometric progression formula,

$$\frac{a}{1-r}.$$

Here $r = -\frac{1}{2}$, so that

$$S = \frac{\frac{1}{2}}{1+\frac{1}{2}} = \frac{1}{3},$$

and α_n tends to $\frac{1}{3}\theta$ as n increases.

I I

Some unsolved problems of modern geometry

CONVEX SETS AND GEOMETRIC INEQUALITIES

The famous historical problems of the previous chapter have all been satisfactorily disposed of. What is left to be done? Most current research in any branch of mathematics is beyond our scope here; but we can perhaps indicate at least some of the kinds of things that are occupying the attention of geometers today.

The field of *convex figures*, brought under investigation (one might better say, invented) only in recent years, is still wide open. A convex plane figure is one that completely contains the straight line segment joining any two points of the figure. Thus the area of Fig. 107A is convex, that of 107B is not. Note that convexity does not necessarily imply curves: every triangle, together with its interior points, comprises a convex set. The definition applies also to convex bodies in 3-space.

The greatest distance between any two points of a plane figure, whether convex or not, is called its diameter. A problem set by Lebesgue in 1914 is still unsolved: to find the figure of least area that will cover (or contain) any

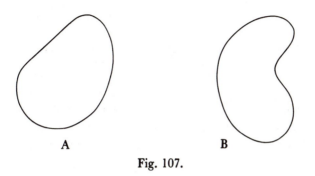

Fig. 107.

figure of unit diameter. The solution was at one time thought to be the shaded area of Fig. 108, formed by a regular hexagon circumscribed around a circle of unit diameter with corners snipped off as indicated. This will not do, however; it fails to cover Fig. 109, which has diameter 1. Even the

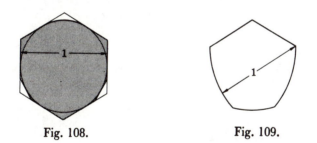

Fig. 108.   **Fig. 109.**

existence of solutions to such problems cannot be taken for granted, as we shall presently see. But in this case a solution is known to exist; the problem is to find (a) the least area, and (b) its shape. It has been proven that the least area lies between .8257 and .8454 square units, inclusive. The upper bound is the area of the hexagon of Fig. 108 with only the two triangles snipped off. The shaded part of Fig. 108 has area .8319, which lies between the two bounds. Even though this figure fails to be a solution, it does not necessarily raise

the lower bound. There *could* be a solution of different shape
that had less area. Although this hardly seems likely, one
cannot be sure without proof.

A *disk* is a circle together with its interior points. Let A and
B be equal disks. If A is cut by a chord into two pieces A_1
and A_2, what is the smallest square that covers A_1, A_2, and
B placed so as not to overlap each other? If h is the height of
the smaller piece of A, then the solution is known to depend
on h and has been completely determined for some h. For
instance, if h is large enough, a configuration like that of
Fig. 110A will do, but for lesser h it is necessary to switch to
Fig. 110B. The transition values of h are not known. A

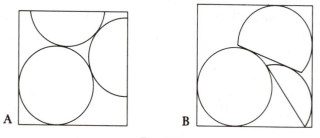

Fig. 110.

formula for the length of the edge of the square in terms of h
is what is wanted here. In three dimensions the problem
may be more difficult: Given two equal spherical balls, one
sliced into two pieces by a plane, what is the smallest cubical
box that will contain these three convex bodies?

If a convex body floats in equilibrium in any orientation,
is it necessarily a sphere? H. T. Croft has suggested some
associated problems: Denote plane sections of a convex body
as follows:

V: those that cut off a certain constant volume;
P: those whose section has a certain constant area;

S: those that cut off a certain constant surface area of the body;

T: those making a constant angle with the tangent planes to the body at all points of the boundary.

Then ask questions by pairing letters; for instance, one could ask, if all sections of type V are also of type S, is the body necessarily a sphere? I do not know whether any of these problems have been solved.

If the edges of a convex polyhedron are all tangent to a sphere of unit radius so as to form a "crate" from which the sphere cannot escape, what is the minimum possible total length of the edges?

MALFATTI'S PROBLEM

Given a piece of marble in the form of a right triangular prism, how can one cut three cylindrical columns from it so as to waste the least possible amount of marble? The wording means not that the cross section is a right triangle, but that the axis of the prism is perpendicular to the base of the prism.

Gianfrancesco Malfatti stated this problem in 1803, and it has attracted the attention of Euclidean geometers from time to time ever since. The problem is equivalent to inscribing three circles in a triangle so that the sum of their areas is a maximum. Malfatti and many others assumed that the three circles should be mutually tangent and each tangent to only two sides of the triangle, as in Fig. 111. The game was, at first, to find a ruler and compass construction of this solution. Many interesting ones were invented, and after a time the problem was considered solved and put on the shelf.

Not until 1930 did anyone notice that the arrangement of Fig. 111 is not always the solution of the original problem.

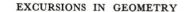

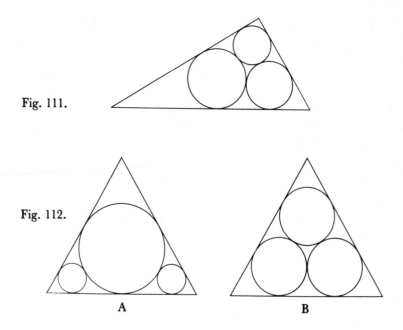

Fig. 111.

Fig. 112.

A   B

Even for the equilateral triangle the circles of Fig. 112A have total area greater than those of Fig. 112B. Then Howard Eves pointed out in 1965 that if the triangle is long and thin, the circles of Fig. 113A have a combined area almost twice as great as those of Fig. 113B, the classical "solution." It is astonishing that this escaped notice for 160 years.

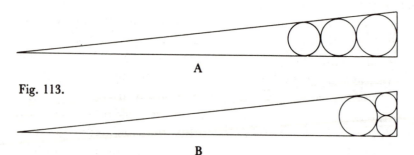

A

Fig. 113.

B

Indeed, the final joke is on the mathematicians: The Malfatti configuration is *never* the solution, no matter what the shape of the triangle! This was announced in 1967 by Michael Goldberg, who showed that the best arrangement is always that of Fig. 112A or 113A. (There is a transition case (Fig. 114) in which the two arrangements provide the same sized circles.) Goldberg's conclusions are based on calculations and graphs; a purely mathematical proof is doubtless difficult and has not yet been produced.

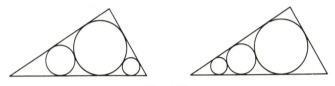

Fig. 114.

THE KAKEYA PROBLEM

If a circle rolls without slipping around the inside of a larger circle, a point on the circumference of the rolling circle traces out a curve. If the fixed circle has diameter exactly three times that of the rolling circle, the curve is called a hypocycloid of three cusps, or *deltoid* (Fig. 115).

In 1917 a Japanese mathematician named S. Kakeya posed the problem: What is the smallest area inside which a line segment of unit length may be turned through 360°? During the next ten years the problem was attacked by many first-class mathematicians without success. Then it came to the attention of A. S. Besicovitch, who published his unexpected solution in 1928.

If the line segment is taken to be the diameter of a circle, it can be rotated (about the center as a pivot) through 360°

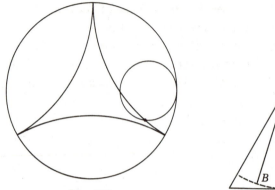

Fig. 115.

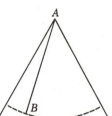

Fig. 116.

without going outside the circle. The circle of unit diameter has radius = $\frac{1}{2}$ and area = $\pi(\frac{1}{2})^2 = \pi/4 = .78...$. But this is not the figure of *least* area. The segment can be turned end for end, by suitable backing and sliding, in an equilateral triangle (Fig. 116). When the segment reaches the end of the indicated arc it must be slid along the side of the triangle and then the next vertex is used as the pivot. This triangle has area = $1/\sqrt{3} = .58...$.

We can do better. The smallest deltoid big enough to permit the turning has an area of $\pi/8$, just half that of the circle. Because it can be shown that this hypocycloid has the property that both ends of the unit tangent line just reach the boundary curve during the entire turning process (Fig. 117), the deltoid seems to be a very "economical" area; and during the decade in which the problem remained unsolved it was generally conjectured that the area could not be reduced below $\pi/8$.

The segment can be turned through 360° in another way, as suggested by Fig. 118. The segment (marked with an

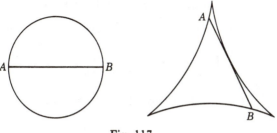

Fig. 117.

arrowhead) is first carried from the position *OP* around the semicircular band, or half-annulus, as far as it will go. It is then backed out to the auxiliary sector *BAC*. Motions in a straight line along the direction of the segment required no area, yet do not violate the statement of the problem, which implies only that the motion be continuous. (Such movements are essential to the ultimate Besicovitch solution.) From *AB* the segment is rotated to *AC* and then brought back along a straight line to the initial position. It has now been turned through 180°, and a second tour will complete the 360° rotation.

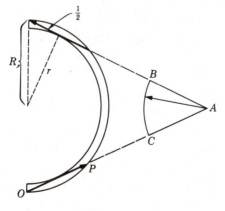

Fig. 118.

It is easy to calculate the area A of the semi-ring of Fig. 118. It is the difference between the areas of two semicircles:

$$A = \frac{\pi R^2}{2} - \frac{\pi r^2}{2} = \frac{\pi}{2}(R^2 - r^2)$$

From the right triangle at the top of the figure, $R^2 = r^2 + (\frac{1}{2})^2$. Therefore

$$A = \frac{\pi}{2}\left[r^2 + (\tfrac{1}{2})^2 - r^2 \right] = \frac{\pi}{8},$$

a constant, no matter how large R and r become, provided they remain just far enough apart to accommodate the unit tangent. Now if we increase these radii, with this proviso, then the point A moves far out to the right and the angle BAC decreases. Thus the area of the sector BAC may be made arbitrarily small. By taking sufficiently large R and r, we can turn the segment through 360° using an area only a very little larger than $\pi/8$, and we can make this "very little" as small as we please.

Evidence acquired by trial and error often serves as a guide in searching for a proof. If different methods seem to lead toward the same result, confidence that we are on the right track is greatly strengthened. But this time we are led astray: two entirely different approaches indicate the same wrong answer, $\pi/8$.

Besicovitch proved that there is *no* least area; that the area in which the segment can be turned can be made arbitrarily small. A segment an inch long can be turned through 360° using less than a thousandth of a square inch of area, or less than a millionth of a square inch, or less than any other area you care to name, however small.

The proof, though long, contains only elementary mathematics. Professor Besicovitch has written a lucid account of it for the Mathematical Association of America (see the Notes). We indicate the general idea of the method.

The segment must be taken through a large number of moves of two types, namely translations and rotations. We have already suggested (Fig. 118) how a translation can be effected using no more than an arbitrarily small area. Suppose the segment must be moved from position *a* to a parallel position *b* (Fig. 119). Two turns through the angle

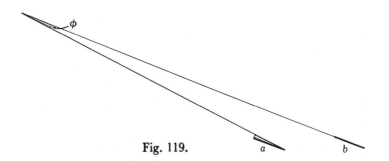

Fig. 119.

φ suffice, and φ can be made as small as you please by carrying the segment for a sufficiently great distance along a line parallel to itself. Thus only the small shaded areas are counted as used area.

Now suppose we wish to rotate the segment through a specified angle, say 30°. One way to do it would be to pivot it within the triangular area shown in Fig. 120A, from *x* through *y* to *z*. If this triangle is bisected along the line *y* and the two halves placed so as to overlap as in Fig. 120B, the segment can now be turned through 30° in an area that is less than before by an amount almost equal to the shaded area. When the segment gets to *y*, it is moved to *y'* by a translation (Fig. 119) using only a tiny area, much less than the shaded area of Fig. 120B. Thus a considerable net reduction has been effected. By a very ingenious sequence of

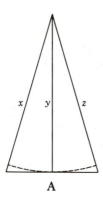 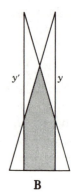

Fig. 120.

repetitions of this idea, Besicovitch was able to show how to use only an area as small as any you care to name in advance.

There are aspects of the problem that remain unsolved. Is there a smallest *simply connected* region in which the turning can be done? The Besicovitch solution employs a highly multiply connected region. A simply connected region has a continuous boundary that does not cross itself; the Deltoid is such a region, but it is not the smallest. The segment can be turned inside a five-pointed curvilinear star of area approximately three-fourths of $\pi/8$. This is done by suitable "backing and filling" within the area indicated in Fig. 121.

Having had such good results with five points, why not try 7, 9, 11, ...? Maybe the limit approached by such a sequence is zero. Unfortunately, it appears not to be zero. Then what is it? And does the area continue to decrease, or is there actually an optimum number of points? That is, is there a specific number N, such that a star of N points has less area than any other star in which the segment can be turned? If so, find N. Are stars of this type the ultimate answer, or must we search among other shapes for smaller simply connected regions? These questions have not been answered.

Modern geometers are concerned with a host of problems, some very different from the few we have been able to mention in this chapter. In addition to the continued study of pure geometry, much work is being done in various spaces (Riemannian, fibre, Grassmann, and others), the geometric aspects of topology, and a wide variety of allied topics. I hope that your interest has been aroused. The Notes give suggestions for further reading. There is literally no end to the trail.

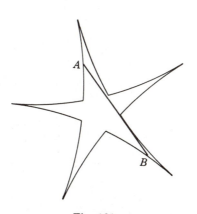

Fig. 121.

Notes

The number refers to the corresponding page of the text

7. *Theorem 1.* An angle inscribed in a circle is measured by half the intercepted arc.

Take first the situation (Fig. 122) where one side *BC*

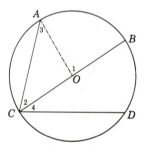

Fig. 122.

of the angle happens to be a diameter. Angle 1 at the center *O* is measured by arc *AB*: the bigger the angle the bigger the arc, in direct proportion. But triangle *COA* is isosceles, and hence ∠1, an exterior angle equal to the sum of the two opposite interior angles ∠2 and

∠3, also equals twice ∠2. Thus ∠2 is measured by half arc *AB*. Now we can do the same for ∠4; and hence finally ∠2+∠4, a general inscribed angle, is measured by half arc *ABD*.

9. Figure 123 is Fig. 3B with construction lines added. Whatever part of the circle (360°) measures ∠1, the rest of it measures ∠3. Remembering to divide by 2 (*half* the intercepted arcs), ∠1+∠3 = 180°. But ∠2+∠3 = 180°. Hence ∠1 = ∠2, and triangles *PCA* and *PBD* have ∠*P* in common and so are similar. Therefore *PA*/*PC* = *PD*/*PB*, and *PA·PB* = *PC·PD*.

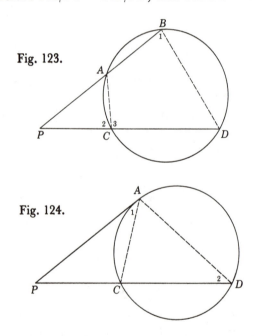

Fig. 123.

Fig. 124.

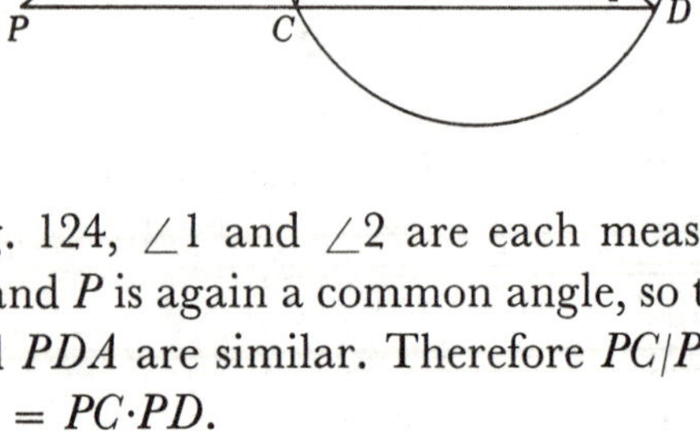

In Fig. 124, ∠1 and ∠2 are each measured by half arc *AC*, and *P* is again a common angle, so that triangles *PAC* and *PDA* are similar. Therefore *PC*/*PA* = *PA*/*PD*, and *PA*² = *PC·PD*.

16. (Why?) Two *right* triangles that share a side and have an acute angle of one equal to the corresponding acute angle of the other are congruent.

25. The strength of the logarithm idea is perhaps better brought out by examples involving roots of numbers. To find the fifth root of a number, one looks up (transforms to) the logarithm of the number, divides that logarithm by 5, and then transforms back by looking up the "anti-log" of the result.

26. (State it.) P and P' are inverse points with respect to a circle if they divide a diameter harmonically.

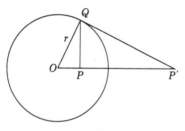

Fig. 125.

28. Proof (IIc): Referring to Fig. 125,

$$\triangle OPQ \text{ is similar to } \triangle OQP'$$

$$\therefore \frac{OP}{r} = \frac{r}{OP'}$$

$$OP \cdot OP' = r^2$$

32. The "trivial" but far-reaching theorem is due to Bolzano: If a function (variable) has a positive value at a and changes continuously to a negative value at b, then somewhere between a and b it must pass through

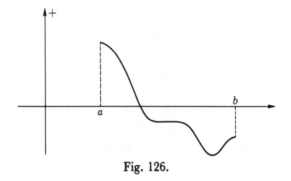

Fig. 126.

zero. The graph (Fig. 126) indicates that $f(x)$ is positive at a and negative at b, and it just can't get there continuously without crossing the X axis somewhere in between. Lest the eye of some horrified mathematician stray to this page, I hasten to assure you (and him) that this is not a proof nor even an accurate statement of Bolzano's Theorem; but it will do for our purposes.

Obvious, isn't it? But *trivial*? No. Observe what it does for us. Think of an arrow (vector) that points

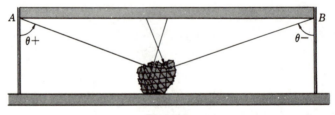

Fig. 127.

from the left-most particle of string in Fig. 127 when it is stretched, to the *same particle* when it is tied around the parcel. Consider the projection of the vector upon the plane of the paper, because we do not care about the component perpendicular to the plane of the paper. Let θ be the angle between that projection and the vertical, measured + to the right of straight down, − to the left of straight down. Straight down is an angle of zero degrees. Now if the tail of the vector is moved continuously along the line occupied by the stretched string from A to B, its head wanders continuously about the parcel, always touching that point of the string that came from the spot where the tail is now. In consequence of the continuity, θ is also a continuous function. It starts with a positive value and ends with a negative value, so it must pass through zero at least once, quite possibly more than once, on its way. When it does, the head of the arrow and the tail of the arrow will both be at the exact same distance from wall A.

36. Here is a hammer-and-tongs proof that the center of a circle does not invert to the center of the image circle: centers are not preserved. (We have a better proof in store.)

In Fig. 25, let $OP = p$, $OQ = q$, etc. The center of circle PQR is then halfway between P and Q, or at a distance $(p+q)/2$ from O. If centers are preserved,

$$\frac{p+q}{2} \cdot \frac{p'+q'}{2} = r^2$$

$$\frac{p+q}{2} \cdot \frac{r^2/p+r^2/q}{2} = r^2$$

$$(p+q)\ (1/p+1/q) = 4$$

$$(p+q)^2 = 4pq$$

Squaring and collecting, this reduces to $(p-q)^2 = 0$, which says that $p = q$. Thus the assumption that centers

are preserved forces the diameter to be zero, or says
that the only such circle is a point.

What if the circle in question is orthogonal to the
circle of inversion? Then the circle is, in a certain sense,
invariant: it transforms to the same circle, although its
individual points move. Even so, its center is not
invariant. A glance at Fig. 21 shows that the center of
circle β is always outside the circle of inversion, and
hence its image will be some point inside the circle of
inversion.

38. Those who have some familiarity with calculus might
like to see a short analytic proof of Theorem 11.

If the transformation takes $P(r,\theta)$ into $P'(\rho,\theta')$, then
$\rho = 1/r$ and $\theta' = \theta$.

$$\frac{d\rho}{dr} = -\frac{1}{r^2}.$$

$$\tan \Psi' = \rho\frac{d\theta}{d\rho} = \rho\frac{d\theta}{dr}\frac{dr}{d\rho} = \frac{1}{r}\frac{d\theta}{dr}(-r^2) = -r\frac{d\theta}{dr} = -\tan \Psi.$$

44. In doing the diameter problem without inversion you
might be tempted to draw the straight lines *AFB*, *BDC*,
and *CEA*, and claim that the problem is now solved
because the altitudes are concurrent. But you can't go
quite so fast. Do you see why not? The question is, are
these three sets of points, for instance, *AFB*, actually
collinear? We can draw *AF* and *FB*, but we cannot
draw *AFB* as a single line without checking this
question.

AO is given as a diameter, so $\angle AFO$ is a right angle
(inscribed in a semicircle). The same applies to $\angle OFB$.
Therefore *AF* and *FB* do indeed form parts of the same
straight line. But the same reasoning *cannot* be applied

to the pieces of *AEC* and *BDE*, because we do not know whether *OC* is a diameter. That is precisely what we have to prove. Draw *AE*, and extend it until it cuts the third circle, say at *X*. Now *OEA* is a right angle, and therefore *OEX* is also. Hence *OX* is a diameter. Do the same for *BD*, letting it intersect the third circle at *Y*: *OY* is likewise a diameter. But there is only one diameter through the point *O*, and thus *X* and *Y* must coincide. We may now call that point of coincidence *C*, and we have that *OC* is a diameter.

Is the problem finished? Why not? Because we still do not know whether *FO* and *OC* are parts of the same straight line. But by this time we have enough to work with. *ABC* is now known to be a triangle, with *BE* and *CF* two altitudes intersecting at *O*. Now *OC* is part of the third altitude because it passes through *O*, and *OF* is part of the same altitude because *AFO* is a right angle. Hence *O*, *F*, and *C* are collinear, and the theorem is finally proved.

It is worth noting that an otherwise excellent book, published recently, contains a rather serious error involving this very problem. Perhaps the moral is that even the best mathematicians sometimes make mistakes. The author starts essentially with our Fig. 32, inverts it, obtains Fig. 31, and then reasons as follows: Because of the orthogonality at *A'*, *B'*, and *C'*, it follows that there is orthogonality at *A*, *B*, and *C*, and therefore we have the theorem that if three circles have a point in common, then the common chord (extended) of each pair is a diameter of the third. That such a theorem is patently false is illustrated by Fig. 128, in which none of the common chords goes anywhere near the center of the third circle. Where did the author go wrong?

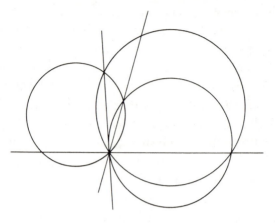

Fig. 128.

49. An alternative simplifying procedure is to shrink all three radii equally until one of the given circles becomes a point, and then perform the inversion with respect to that point.

50. (Why?) A and B intersect only at O; hence their images, A' and B', intersect only at infinity, which says that they are parallel.

50. What is the significance of the second one? Y would yield another solution. In this particular drawing, because of the orientation D' would then touch all three of A', B', and C' inside the circle of inversion. D would therefore touch each of A, B, and C outside the circle of inversion, and would in fact lead to the solution that surrounds all three.

52. (The procedure of Fig. 21.) Any circle centered on the radical axis of the α circles which is orthogonal to one α circle is automatically orthogonal to all of them (Fig. 15).

52. Jacob Steiner, 1796–1867.

54. It is sloppy mathematics to talk about fractional parts

of circles without defining what is meant; but the idea is clear. For properly presented analytic equations covering the requirements for the closing of a Steiner chain after 1 or n circuits, see *Geometry Revisited*, by H. S. M. Coxeter and S. L. Greitzer (Random House & L. W. Singer, 1967), p. 125.

56. Answer to Query: By taking $d_2 - d_1 =$ the same constant.
57. Theorem 14 means that, for whatever of the infinite number of positions the Steiner chain may assume for two given base circles, the circles of the chain always have their centers on the same ellipse.

We now have an elegant proof that centers of circles are not preserved under inversion. For if they were, the circle on which the centers of the movable circles all lie in Fig. 40 would have to become the ellipse on which they lie in Fig. 39, an impossibility because a circle never inverts to an ellipse.

59. If the fixed circles of Theorem 14 Generalized have equal radii, the last equation of the proof reduces to $PO_2 = PO_1$. All such points P lie on the perpendicular bisector of the line O_1O_2. A straight line is a "degenerate" conic.
59. Is there a parabolic case? Yes, when one of the two fixed circles is a straight line (circle of infinite radius).
60. 3-space is just a name for the ordinary space of three dimensions in which we (think we) live.
60. Soddy's Hexlet: *Nature*, Vol. 139 (1937): Soddy, pp. 77, 154, 252; Frank Morley, p. 72; Thorold Gosset, p. 251.
61. That six pennies will group snugly around a seventh is a consequence of the fact that six equilateral triangles of the same size form a hexagon. All seven circles have radii equal to one-half the side of a triangle; six of them are centered at the vertices of the hexagon and the seventh at its center.

68. The proof using the hyperboloid of revolution was devised by J. H. Cadwell of the British Royal Aircraft Establishment in 1968 and communicated to me by letter. It is published here with his kind permission.

     The last statement of the proof (that the plane section of a hyperboloid of revolution is a conic) is standard analytic geometry. For the equation of a quadric surface is always of second degree, and that of a plane is linear. If we eliminate one variable between these two equations, the result is a quadratic in two variables. This is a right cylinder, not circular, but based on a conic. But the parallel projection of a conic upon a plane is another conic.

74. (Then where would it go?) After a second reflection, back through the first focus—and so on. It quickly approaches a path of oscillation back and forth along the ellipse's major axis. See *Mathematical Snapshots*, by Hugo Steinhaus (Oxford University Press, New York, 1969), p. 239.

81. (Why?) $Q_1 Q_2$ is always measured along an element of the cone (a straight line through the vertex). But any two plane sections cutting a cone in circles C_1 and C_2 are perpendicular to the axis of the cone, and hence intercept equal segments on all the elements.

81. Consider the cone of light rays that just graze the beach ball. Then the ball is the lesser Dandelin sphere, tangent to the elliptical shadow at its focus.

81. The Dandelin proof for the parabola can be found in the author's *A Calculus Notebook* (Prindle, Weber & Schmidt, Boston, 1968), p. 11. The proof for the hyperbola is presented on page 9 of Hilbert and Cohn-Vossen's *Geometry and the Imagination* (Chelsea, New York, 1952).

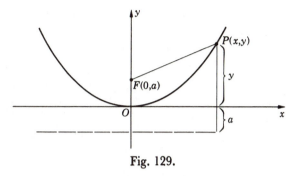

Fig. 129.

82. Let the focus of the parabola be the point $(0,a)$ and the directrix be the line $y = -a$ (Fig. 129). Then by definition, $PF = PD$. By the Pythagorean distance formula, the distance between the points $P(x,y)$ and $F(0,a)$ is $PF = \sqrt{(x-0)^2 + (y-a)^2}$. On the other hand, from the figure, $PD = y+a$. Equating, and squaring both sides:

$$(x-0)^2 + (y-a)^2 = (y+a)^2$$
$$x^2 + y^2 - 2ay + a^2 = y^2 + 2ay + a^2$$
$$x^2 = 4ay$$

$4a$ is a constant which we can call k to obtain the desired equation.

83. The parabola is the limiting case of the ellipse as the cutting plane of Fig. 59 approaches a position parallel to an element of the cone. This can be stated in terms of the *eccentricity* of the ellipse, defined as $e = c/a$, where c is the distance from the center to a focus and a is the distance from the center to a vertex (end) of the ellipse.

   Now suppose that one focus and the corresponding vertex were to remain fixed while the other focus and vertex receded indefinitely far. Because the center would recede too, c and a would tend toward equality with each other. In fact the eccentricity of a parabola (defined differently) is 1.

You might guess, then, that an ellipse with eccentricity .99 must be exceedingly long and thin; for if the eccentricity is 1 it is "infinitely long." One more guess gone wrong. Those who know the "ellipse triangle" will recognize

$$
\begin{aligned}
b^2 &= a^2 - c^2 \\
&= 100^2 - 99^2 \\
&= (100+99)\,(100-99) \\
&= 199 \\
b &= 14.1
\end{aligned}
$$

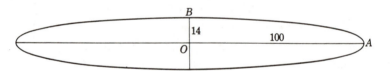

Fig. 130.

Thus the proportions of this ellipse are about 7 to 1 (Fig. 130). Note that the focus is 99/100 of the way from O to A, almost impossible to show on a drawing of this scale. In most textbooks you will find the foci plotted too far away from the vertices of ellipses.

The orbits of the planets, though elliptical, have small eccentricities. Mercury's orbit, the most eccentric, has $e = 1/5$, making the major and minor axes differ by about 2 per cent. Such an ellipse could probably not be distinguished from a circle by the keenest eye. The earth's orbit, with eccentricity $= 1/60$, is much closer to a perfect circle than Mercury's, the major and minor axes differing by about one part in 8000.

106. Well? Consider the set of all hexagons inscribable in the given conic. Let Class A contain all those that have opposite sides parallel. There are infinitely many members of Class A, and they take a wide variety of shapes; but they do not exhaust the possibilities. There are also infinitely many members of Class B, whose opposite sides are not parallel. What we have shown is that Class A hexagons project into illustrations of Pascal's Theorem. We know that Class B hexagons do project into hexagons inscribed in conics—but our "proof" says not a word about whether they comply with Pascal's Theorem when they get there, and we might even guess that they would not in the absence of further information.

If you are not convinced of the fallacy of the "proof," try it with octagons. If this line of attack were valid, there ought to be a Pascal's Theorem for octagons—but there isn't. The intersections of opposite pairs of sides of Class B octagons are *not* collinear.

For eleven other proofs, see "Various proofs of Pascal's Theorem," Kaidy Tan, *Mathematics Magazine*, Vol. 38 (1965), p. 22.

107. Strictly, *triangle* should read *trilateral* in the right-hand column.

108. The postulates for plane projective geometry are taken from Vol. I of *A Survey of Geometry*, by Howard Eves (Allyn and Bacon, Boston, 1963), p. 418.

110.

PAPPUS' THEOREM	DUAL OF PAPPUS'
The 3 points, lying on the 3 pairs of opposite sides of a hexagon whose vertices lie alternately on 2 lines, lie on one line.	The 3 lines, lying on the 3 pairs of opposite vertices of a hexagon whose sides lie alternately on 2 points, lie on one point.

PASCAL'S THEOREM

(*Generalization of Pappus'*)

The 3 points, lying on the 3 pairs
of opposite sides of a hexagon
whose vertices lie on a conic,
lie on one line.

BRIANCHON'S THEOREM

(*Generalization of Pappus' Dual*)

The 3 lines, lying on the 3 pairs
of opposite vertices of a hexagon
whose sides lie on a conic,
lie on one point.

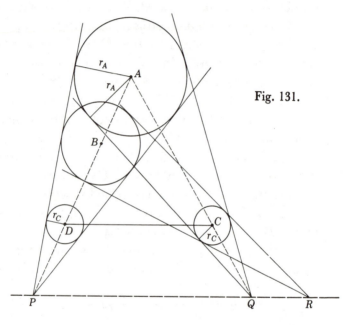

Fig. 131.

115. The three-circle problem. In Fig. 131, D is the circle
of radius r_C that is tangent to both the external
tangents of A and B.

$$\frac{PD}{PA} = \frac{r_C}{r_A} = \frac{QC}{QA}$$

Hence DC is parallel to PQ. In the same way, DC is
parallel to PR. Therefore segment PQ is part of line
PR.

This proof and the three-dimensional version are given by L. A. Graham in *Ingenious Mathematical Problems and Methods* (Dover, New York, 1959), No. 62. Coxeter proves the theorem in a much more elaborate way, for three circles each of which is entirely external to the other two. "The Problem of Apollonius," H. S. M. Coxeter, *American Mathematical Monthly*, Vol. 75 (1968), p. 14.

118. (Why?) Each perpendicular bisector is the locus of points equidistant from some pair of vertices; therefore any two of them intersect at *P*.

119. Daniel Pedoe, on page one of his book entitled *Circles* (Pergamon, New York, 1957), calls the nine-point circle "the first really exciting one to appear in any course on elementary geometry."

120. The triangle problem is No. E-2124, *American Mathematical Monthly*, Vol. 75 (1968), p. 899.

Your diagram may not look quite like Fig. 132, but the proof will be the same except possibly for some sign changes. We will show that $AP = AQ$ because they are corresponding parts of congruent triangles; and then that $\gamma = 90°$ because $\alpha + \beta = 90°$, and the three interior angles of a triangle must total 180°.

First observe that $DC = CB$ and $DQ = CF = CA$. Also $\angle x$ at $D = \angle x$ at C: the two sides of the one are respectively perpendicular to the two sides of the other. Hence $\triangle DQC$ is congruent to $\triangle CAB$, by side-angle-side. In exactly the same way, $\triangle BPE$ is congruent to $\triangle CAB$. From these we have the corresponding sides $QC = AB$ and $BP = CA$. But $\angle ACQ = x + 90° + y = \angle PBA$. Hence triangles ACQ and ABP are congruent by side-angle-side. Therefore $AP = AQ$.

Next, the exterior angle α is equal to the sum of the

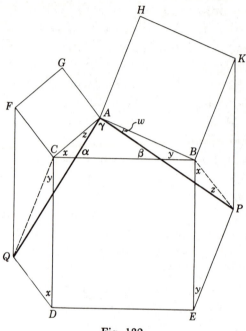

Fig. 132.

two opposite interior angles $x+z$. Likewise $\beta = w+y$.
Adding, $\alpha+\beta = x+z+w+y$. But in triangle ABP, all
the interior angles total 180°:

$$x+z+w+y+90° = 180°$$

or,

$$x+z+w+y = 90°.$$

Substituting,

$$\alpha+\beta = 90°$$

$$\therefore \quad \gamma = 90°.$$

121. For further reading, see *Introduction to Geometry*, by
H. S. M. Coxeter (John Wiley & Sons, Inc., New
York, 1961); *A Survey of Geometry*, by Howard Eves
(Allyn & Bacon, Inc., Boston, 1963, especially Vol. I);

College Geometry, by N. A. Court (2nd Ed., Barnes & Noble, New York, 1952).

136. To find π is to find $\sqrt{\pi}$, because square roots are constructible.

136. An excellent treatment of this whole topic, with all proofs carefully completed, is to be found in *What is Mathematics?* by R. Courant and H. Robbins (Oxford University Press, 1941), pp. 117–40. See also *Number, The Language of Science*, by Tobias Dantzig (Macmillan, New York, 1954), p. 316.

136. Lindemann's 1882 proof of the transcendence of π, somewhat simplified but still not easy, is presented in *Famous Problems*, by Felix Klein (Chelsea, New York, 1955).

137. Duplication of the cube. We go too fast here. It is also necessary to be sure that the cubic equation $x^3 - 2 = 0$ is irreducible.

137. Gauss's proof appeared in his *Disquisitiones Arithmeticae* in 1801, a book that has only recently been translated into English and published for the first time in 1966 by Yale University Press. The converse theorem, namely that no other regular polygons are constructible, has usually been credited also to Gauss, but N. D. Kazarinoff has pointed out that the first proof was actually published by Pierre L. Wantzel in 1837. See *American Mathematical Monthly*, Vol. 75 (1968), p. 647.

139. 1. $\angle 1 = \angle 2$, alternate interior angles.

      $\angle 2 = \angle 3$, base angles of an isosceles triangle.

      $\angle 3 = 2\angle 4$, exterior angle = sum of two opposite interior angles.

      $\angle 4 = \angle 5$, interior-exterior angles.

140. 3. The trisection is incidental to a curve called the

Conchoid of Nichomedes. See *A Book of Curves*, by E. H. Lockwood (Cambridge University Press, 1961), p. 127.

142. Henri Lebesgue, 1875–1941.

143. "A solution is known to exist." Page 18 (first footnote) and p. 100 of *Convex Figures*, by I. M. Yaglom and V. G. Boltyanskii (Holt, Rinehart & Winston, New York, 1961). This volume contains many unsolved problems and hundreds of pertinent references.

143. The arcs of Fig. 109 are all of radius 1, and are centered at the three top vertices of the hexagon. This counter-example was submitted by B. Abel de Valcourt of the University of Minnesota. The upper and lower bounds on the area are due to J. Pal:

$$.8257 \ldots = \frac{\pi}{8} + \frac{\sqrt{3}}{4}$$

$$.8454 \ldots = \frac{2}{3}(3 - \sqrt{3})$$

These and other data can be found in "Borsuk's Problem and Related Questions," by Branko Gruenbaum, in *Convexity—Proceedings of Symposia in Pure Mathematics*, Vol. 7, p. 274 (American Mathematical Society, 1963).

144. The minimum square is part of Problem E-1924, solution by Michael Goldberg, *American Mathematical Monthly*, Vol. 75 (1968), p. 195.

145. The question on the crated sphere is asked by H. S. M. Coxeter on p. 495 of *Convexity* (see note to page 143, above). See also G. C. Shephard's "A Sphere in a Crate", *Journal London Mathematical Society*, Vol. 40 (1965), p. 433.

146. If the equilateral triangle of Fig. 112 has altitude 3, so that the radius of the largest inscribed circle is 1, then the total area of the circles of 112A is 1.222π, as opposed to 1.206π for 112B—a little over 1 per cent more for the non-Malfatti arrangement. Goldberg shows that the Malfatti solution comes closest to being the correct one in this equilateral case. "On the Original Malfatti Problem," by Michael Goldberg, *Mathematics Magazine*, Vol. 40 (1967), p. 241.

146. Howard Eves, *A Survey of Geometry* (Allyn & Bacon, Boston, 1965), Vol. 2, p. 245.

147. Part of this description of the Kakeya problem appeared in the author's *A Calculus Notebook* (Prindle, Weber & Schmidt, Boston, 1968), and is reproduced here with the permission of the publishers.

151. "The Kakeya Problem," by A. S. Besicovitch, *American Mathematical Monthly*, Vol. 70 (1963), p. 697.

152. The five-pointed star was produced by R. J. Walker of Cornell University in 1952. (See "Editorial Note," *American Mathematical Monthly*, Vol. 71 (1964), p. 516.) Walker used arcs of circles rather than a hypocycloid for his star, whose area was calculated by approximation methods.

Finding a workable formula for area is not always as easy as it sounds, even for figures bounded by straight lines; and in three dimensions the situation is still more difficult. For example, no method is known for computing the volume of a general convex polyhedron. *Proceedings of CUPM Geometry Conference*, Mathematical Association of America, No. 16 (1967), p. 21.

Index

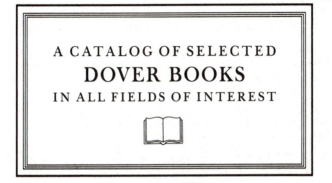

A CATALOG OF SELECTED

DOVER BOOKS

IN ALL FIELDS OF INTEREST

A CATALOG OF SELECTED DOVER
BOOKS IN ALL FIELDS OF INTEREST

CONCERNING THE SPIRITUAL IN ART, Wassily Kandinsky. Pioneering work by father of abstract art. Thoughts on color theory, nature of art. Analysis of earlier masters. 12 illustrations. 80pp. of text. 5⅜ x 8½.　　　23411-8 Pa. $3.95

ANIMALS: 1,419 Copyright-Free Illustrations of Mammals, Birds, Fish, Insects, etc., Jim Harter (ed.). Clear wood engravings present, in extremely lifelike poses, over 1,000 species of animals. One of the most extensive pictorial sourcebooks of its kind. Captions. Index. 284pp. 9 x 12.　　　23766-4 Pa. $12.95

CELTIC ART: The Methods of Construction, George Bain. Simple geometric techniques for making Celtic interlacements, spirals, Kells-type initials, animals, humans, etc. Over 500 illustrations. 160pp. 9 x 12. (USO)　　　22923-8 Pa. $9.95

AN ATLAS OF ANATOMY FOR ARTISTS, Fritz Schider. Most thorough reference work on art anatomy in the world. Hundreds of illustrations, including selections from works by Vesalius, Leonardo, Goya, Ingres, Michelangelo, others. 593 illustrations. 192pp. 7⅛ x 10¼.　　　20241-0 Pa. $9 95

CELTIC HAND STROKE-BY-STROKE (Irish Half-Uncial from "The Book of Kells"): An Arthur Baker Calligraphy Manual, Arthur Baker. Complete guide to creating each letter of the alphabet in distinctive Celtic manner. Covers hand position, strokes, pens, inks, paper, more. Illustrated. 48pp. 8¼ x 11.　　　24336-2 Pa. $3.95

EASY ORIGAMI, John Montroll. Charming collection of 32 projects (hat, cup, pelican, piano, swan, many more) specially designed for the novice origami hobbyist. Clearly illustrated easy-to-follow instructions insure that even beginning papercrafters will achieve successful results. 48pp. 8¼ x 11.　　　27298-2 Pa. $2.95

THE COMPLETE BOOK OF BIRDHOUSE CONSTRUCTION FOR WOODWORKERS, Scott D. Campbell. Detailed instructions, illustrations, tables. Also data on bird habitat and instinct patterns. Bibliography. 3 tables. 63 illustrations in 15 figures. 48pp. 5¼ x 8½.　　　24407-5 Pa. $2.50

BLOOMINGDALE'S ILLUSTRATED 1886 CATALOG: Fashions, Dry Goods and Housewares, Bloomingdale Brothers. Famed merchants' extremely rare catalog depicting about 1,700 products: clothing, housewares, firearms, dry goods, jewelry, more. Invaluable for dating, identifying vintage items. Also, copyright-free graphics for artists, designers. Co-published with Henry Ford Museum & Greenfield Village. 160pp. 8¼ x 11.　　　25780-0 Pa. $9.95

HISTORIC COSTUME IN PICTURES, Braun & Schneider. Over 1,450 costumed figures in clearly detailed engravings–from dawn of civilization to end of 19th century. Captions. Many folk costumes. 256pp. 8⅜ x 11¾.　　　23150-X Pa. $12.95

FRANK LLOYD WRIGHT'S HOLLYHOCK HOUSE, Donald Hoffmann. Lavishly illustrated, carefully documented study of one of Wright's most controversial residential designs. Over 120 photographs, floor plans, elevations, etc. Detailed perceptive text by noted Wright scholar. Index. 128pp. 9¼ x 10¾. 27133-1 Pa. $11.95

THE MALE AND FEMALE FIGURE IN MOTION: 60 Classic Photographic Sequences, Eadweard Muybridge. 60 true-action photographs of men and women walking, running, climbing, bending, turning, etc., reproduced from rare 19th-century masterpiece. vi + 121pp. 9 x 12. 24745-7 Pa. $10.95

1001 QUESTIONS ANSWERED ABOUT THE SEASHORE, N. J. Berrill and Jacquelyn Berrill. Queries answered about dolphins, sea snails, sponges, starfish, fishes, shore birds, many others. Covers appearance, breeding, growth, feeding, much more. 305pp. 5¼ x 8¼. 23366-9 Pa. $8.95

GUIDE TO OWL WATCHING IN NORTH AMERICA, Donald S. Heintzelman. Superb guide offers complete data and descriptions of 19 species: barn owl, screech owl, snowy owl, many more. Expert coverage of owl-watching equipment, conservation, migrations and invasions, etc. Guide to observing sites. 84 illustrations. xiii + 193pp. 5⅜ x 8½. 27344-X Pa. $8.95

MEDICINAL AND OTHER USES OF NORTH AMERICAN PLANTS: A Historical Survey with Special Reference to the Eastern Indian Tribes, Charlotte Erichsen-Brown. Chronological historical citations document 500 years of usage of plants, trees, shrubs native to eastern Canada, northeastern U.S. Also complete identifying information. 343 illustrations. 544pp. 6½ x 9¼. 25951-X Pa. $12.95

STORYBOOK MAZES, Dave Phillips. 23 stories and mazes on two-page spreads: Wizard of Oz, Treasure Island, Robin Hood, etc. Solutions. 64pp. 8¼ x 11. 23628-5 Pa. $2.95

NEGRO FOLK MUSIC, U.S.A., Harold Courlander. Noted folklorist's scholarly yet readable analysis of rich and varied musical tradition. Includes authentic versions of over 40 folk songs. Valuable bibliography and discography. xi + 324pp. 5⅜ x 8½. 27350-4 Pa. $7.95

MOVIE-STAR PORTRAITS OF THE FORTIES, John Kobal (ed.). 163 glamor, studio photos of 106 stars of the 1940s: Rita Hayworth, Ava Gardner, Marlon Brando, Clark Gable, many more. 176pp. 8⅜ x 11¼. 23546-7 Pa. $12.95

BENCHLEY LOST AND FOUND, Robert Benchley. Finest humor from early 30s, about pet peeves, child psychologists, post office and others. Mostly unavailable elsewhere. 73 illustrations by Peter Arno and others. 183pp. 5⅜ x 8½. 22410-4 Pa. $6.95

YEKL and THE IMPORTED BRIDEGROOM AND OTHER STORIES OF YIDDISH NEW YORK, Abraham Cahan. Film Hester Street based on Yekl (1896). Novel, other stories among first about Jewish immigrants on N.Y.'s East Side. 240pp. 5⅜ x 8½. 22427-9 Pa. $6.95

SELECTED POEMS, Walt Whitman. Generous sampling from *Leaves of Grass*. Twenty-four poems include "I Hear America Singing," "Song of the Open Road," "I Sing the Body Electric," "When Lilacs Last in the Dooryard Bloom'd," "O Captain! My Captain!"–all reprinted from an authoritative edition. Lists of titles and first lines. 128pp. 5$\frac{5}{16}$ x 8¼. 26878-0 Pa. $1.00

THE BEST TALES OF HOFFMANN, E. T. A. Hoffmann. 10 of Hoffmann's most important stories: "Nutcracker and the King of Mice," "The Golden Flowerpot," etc. 458pp. 5⅜ x 8½. 21793-0 Pa. $9.95

FROM FETISH TO GOD IN ANCIENT EGYPT, E. A. Wallis Budge. Rich detailed survey of Egyptian conception of "God" and gods, magic, cult of animals, Osiris, more. Also, superb English translations of hymns and legends. 240 illustrations. 545pp. 5⅜ x 8½. 25803-3 Pa. $11.95

FRENCH STORIES/CONTES FRANÇAIS: A Dual-Language Book, Wallace Fowlie. Ten stories by French masters, Voltaire to Camus: "Micromegas" by Voltaire; "The Atheist's Mass" by Balzac; "Minuet" by de Maupassant; "The Guest" by Camus, six more. Excellent English translations on facing pages. Also French-English vocabulary list, exercises, more. 352pp. 5⅜ x 8½. 26443-2 Pa. $8.95

CHICAGO AT THE TURN OF THE CENTURY IN PHOTOGRAPHS: 122 Historic Views from the Collections of the Chicago Historical Society, Larry A. Viskochil. Rare large-format prints offer detailed views of City Hall, State Street, the Loop, Hull House, Union Station, many other landmarks, circa 1904-1913. Introduction. Captions. Maps. 144pp. 9⅜ x 12¼. 24656-6 Pa. $12.95

OLD BROOKLYN IN EARLY PHOTOGRAPHS, 1865-1929, William Lee Younger. Luna Park, Gravesend race track, construction of Grand Army Plaza, moving of Hotel Brighton, etc. 157 previously unpublished photographs. 165pp. 8⅞ x 11¾. 23587-4 Pa. $13.95

THE MYTHS OF THE NORTH AMERICAN INDIANS, Lewis Spence. Rich anthology of the myths and legends of the Algonquins, Iroquois, Pawnees and Sioux, prefaced by an extensive historical and ethnological commentary. 36 illustrations. 480pp. 5⅜ x 8½. 25967-6 Pa. $8.95

AN ENCYCLOPEDIA OF BATTLES: Accounts of Over 1,560 Battles from 1479 B.C. to the Present, David Eggenberger. Essential details of every major battle in recorded history from the first battle of Megiddo in 1479 B.C. to Grenada in 1984. List of Battle Maps. New Appendix covering the years 1967-1984. Index. 99 illustrations. 544pp. 6½ x 9¼. 24913-1 Pa. $14.95

SAILING ALONE AROUND THE WORLD, Captain Joshua Slocum. First man to sail around the world, alone, in small boat. One of great feats of seamanship told in delightful manner. 67 illustrations. 294pp. 5⅜ x 8½. 20326-3 Pa. $5.95

ANARCHISM AND OTHER ESSAYS, Emma Goldman. Powerful, penetrating, prophetic essays on direct action, role of minorities, prison reform, puritan hypocrisy, violence, etc. 271pp. 5⅜ x 8½. 22484-8 Pa. $6.95

MYTHS OF THE HINDUS AND BUDDHISTS, Ananda K. Coomaraswamy and Sister Nivedita. Great stories of the epics; deeds of Krishna, Shiva, taken from puranas, Vedas, folk tales; etc. 32 illustrations. 400pp. 5⅜ x 8½. 21759-0 Pa. $10.95

BEYOND PSYCHOLOGY, Otto Rank. Fear of death, desire of immortality, nature of sexuality, social organization, creativity, according to Rankian system. 291pp. 5⅜ x 8½. 20485-5 Pa. $8.95

A THEOLOGICO-POLITICAL TREATISE, Benedict Spinoza. Also contains unfinished Political Treatise. Great classic on religious liberty, theory of government on common consent. R. Elwes translation. Total of 421pp. 5⅜ x 8½. 20249-6 Pa. $9.95

THE INFLUENCE OF SEA POWER UPON HISTORY, 1660–1783, A. T. Mahan. Influential classic of naval history and tactics still used as text in war colleges. First paperback edition. 4 maps. 24 battle plans. 640pp. 5⅜ x 8½. 25509-3 Pa. $12.95

THE STORY OF THE TITANIC AS TOLD BY ITS SURVIVORS, Jack Winocour (ed.). What it was really like. Panic, despair, shocking inefficiency, and a little heroism. More thrilling than any fictional account. 26 illustrations. 320pp. 5⅜ x 8½. 20610-6 Pa. $8.95

FAIRY AND FOLK TALES OF THE IRISH PEASANTRY, William Butler Yeats (ed.). Treasury of 64 tales from the twilight world of Celtic myth and legend: "The Soul Cages," "The Kildare Pooka," "King O'Toole and his Goose," many more. Introduction and Notes by W. B. Yeats. 352pp. 5⅜ x 8½. 26941-8 Pa. $8.95

BUDDHIST MAHAYANA TEXTS, E. B. Cowell and Others (eds.). Superb, accurate translations of basic documents in Mahayana Buddhism, highly important in history of religions. The Buddha-karita of Asvaghosha, Larger Sukhavativyuha, more. 448pp. 5⅜ x 8½. 25552-2 Pa. $9.95

ONE TWO THREE . . . INFINITY: Facts and Speculations of Science, George Gamow. Great physicist's fascinating, readable overview of contemporary science: number theory, relativity, fourth dimension, entropy, genes, atomic structure, much more. 128 illustrations. Index. 352pp. 5⅜ x 8½. 25664-2 Pa. $8.95

ENGINEERING IN HISTORY, Richard Shelton Kirby, et al. Broad, nontechnical survey of history's major technological advances: birth of Greek science, industrial revolution, electricity and applied science, 20th-century automation, much more. 181 illustrations. ". . . excellent . . ."–*Isis*. Bibliography. vii + 530pp. 5⅜ x 8¼. 26412-2 Pa. $14.95

DALÍ ON MODERN ART: The Cuckolds of Antiquated Modern Art, Salvador Dalí. Influential painter skewers modern art and its practitioners. Outrageous evaluations of Picasso, Cézanne, Turner, more. 15 renderings of paintings discussed. 44 calligraphic decorations by Dalí. 96pp. 5⅜ x 8½. (USO) 29220-7 Pa. $4.95

ANTIQUE PLAYING CARDS: A Pictorial History, Henry René D'Allemagne. Over 900 elaborate, decorative images from rare playing cards (14th–20th centuries): Bacchus, death, dancing dogs, hunting scenes, royal coats of arms, players cheating, much more. 96pp. 9¼ x 12¼. 29265-7 Pa. $11.95

MAKING FURNITURE MASTERPIECES: 30 Projects with Measured Drawings, Franklin H. Gottshall. Step-by-step instructions, illustrations for constructing handsome, useful pieces, among them a Sheraton desk, Chippendale chair, Spanish desk, Queen Anne table and a William and Mary dressing mirror. 224pp. 8⅛ x 11¼. 29338-6 Pa. $13.95

THE FOSSIL BOOK: A Record of Prehistoric Life, Patricia V. Rich et al. Profusely illustrated definitive guide covers everything from single-celled organisms and dinosaurs to birds and mammals and the interplay between climate and man. Over 1,500 illustrations. 760pp. 7½ x 10⅛. 29371-8 Pa. $29.95

Prices subject to change without notice.

Available at your book dealer or write for free catalog to Dept. GI, Dover Publications, Inc., 31 East 2nd St., Mineola, N.Y. 11501. Dover publishes more than 500 books each year on science, elementary and advanced mathematics, biology, music, art, literary history, social sciences and other areas.